2

Lipid Analysis

Lipid Analysis

Isolation, separation, identification and structural
analysis of lipids

WILLIAM W. CHRISTIE

Hannah Research Institute, Ayr, Scotland

PERGAMON PRESS
OXFORD · NEW YORK · TORONTO
SYDNEY · BRAUNSCHWEIG

Pergamon Press Ltd., Headington Hill Hall, Oxford
Pergamon Press Inc., Maxwell House, Fairview Park, Elmsford
New York 10523
Pergamon of Canada Ltd., 207 Queen's Quay West, Toronto 1
Pergamon Press (Aust.) Pty. Ltd., 19a Boundary Street
Rushcutters Bay, N.S.W. 2011, Australia
Vieweg & Sohn GmbH, Burgplatz 1, Braunschweig

First edition 1973

Library of Congress Cataloging in Publication Data

Christie, William W.
Lipid analysis.
Includes bibliographical references.

1. Lipids—Analysis. 1. Title.
QP751.C49 1973   574.1'9247   73–8946
ISBN 0–08–017753–0

*Printed in Great Britain by
Billing & Sons Limited, Guildford and London*

Contents

Preface

TWENTY years ago, lipids were considered to be oily intractable substances that could be separated into simpler components only with great difficulty and they were studied by a comparatively limited number of painstaking devotees. The development of chromatographic techniques, particularly gas–liquid chromatography and thin-layer chromatography, together with advances in spectroscopy, have led to an explosive growth of interest in these compounds and have revolutionised our knowledge of the role that lipids play in the structure and function of cell membranes, as essential dietary components and in numerous biological processes. A great number of specialist journals and review articles are now published at regular intervals and in this book I have attempted to critically examine the literature and bring together in a systematic manner the best of the procedures that have been developed for separating, identifying and determining lipid classes and their component parts. Many of these methods were evaluated at the same time in my own laboratory. I hope that newcomers to the subject will find it a useful guide through the potential complexities of lipid analysis and that experts in the field will find it a valuable reference work.

Acknowledgements

I AM grateful to Professor J. A. F. Rook, Director of the Hannah Research Institute, and to the Department of Agriculture and Fisheries for Scotland for permission to write this book. My thanks are due also to Dr. W. R. Morrison of the University of Strathclyde and Drs. J. H. Moore and R. C. Noble of the Hannah Research Institute for reading critically the first draft; their comments led to many improvements. Miss M. L. Hunter gave considerable assistance in testing some of the procedures described and in checking the final text, and Miss Elizabeth Atkinson carefully and patiently typed two drafts of the complete manuscript. I gladly acknowledge the assistance of the authors and journals cited at points in the text in permitting me to reproduce certain figures. Finally, I must thank my wife Norma for her patience and encouragement through many long evenings while this book was in preparation.

W. W. CHRISTIE

CHAPTER 1

The Structure, Chemistry and Occurrence of Lipids

A. Introduction

The term LIPID is often used loosely to denote a wide variety of natural products including fatty acids and their derivatives, steroids, terpenes, carotenoids and bile acids, which have in common a ready solubility in organic solvents such as diethyl ether, hexane, benzene, chloroform or methanol. A more specific definition is to be preferred and the term is nowadays generally restricted to fatty acids and their derivatives or metabolites. It is in this sense that the term is used in this book.

The principal lipid classes consist of fatty acid (long-chain aliphatic monocarboxylic acid) moieties linked by an ester bond to an alcohol, principally the trihydric alcohol glycerol, or by amide bonds to long-chain bases. Also, they may contain phosphoric acid, organic bases, sugars and more complex components that can be liberated by various hydrolytic procedures. Lipids may be subdivided into two broad classes—"simple", which contain one or two of these hydrolysis products per mole, and "complex", which contain three or more types of hydrolysis product per mole. The terms "neutral" and "polar" respectively are used more frequently to define these classes, but are less precise and may occasionally be ambiguous; for example, unesterified fatty acids are normally classed as neutral lipids despite the presence of the free carboxyl group.

A complete analysis of the lipids from a given source, therefore, involves separation of the lipid mixture into simpler types according to the number and nature of the various constituent parts, the identification and estimation of each of these and finally determination of the absolute amount of each lipid type. Before progressing to this, however,

1

a knowledge of the structure, chemistry and occurrence of the principal known lipids and their constituents is necessary.

B. The Fatty Acids

The common fatty acids of plant and animal origin contain even numbers of carbon atoms (4–24) in straight chains with a terminal carboxyl group and may be fully saturated or contain one, two or more (up to six) double bonds, which generally but not always have a *cis*-configuration. Fatty acids of animal origin are comparatively simple in structure and can be subdivided into well-defined families. Plant fatty acids, on the other hand, may be more complex and can contain a variety of other functional groups including acetylenic bonds, epoxyl, hydroxyl or keto groups and cyclopropene rings. Bacterial fatty acids usually consist of simpler saturated and monoenoic components but may also contain odd-numbered, branched-chain and cyclo-propane acids. Very complex high molecular weight acids, the mycolic acids, have been found in certain bacterial species.

1. *Saturated fatty acids*

The commonest saturated fatty acids are straight-chain even-numbered acids containing 14–20 carbon atoms, although all the possible odd and even-numbered homologues with 2–30 or more carbon atoms have been found in nature. They are named systematically from the saturated hydrocarbon with the same number of carbon atoms, the final -*e* being changed to -*oic*. For example, the acid with sixteen carbon atoms and structural formula

$$CH_3 \cdot (CH_2)_{14} \cdot COOH$$

is correctly termed hexadecanoic acid, although it also has a trivial name hallowed by common usage, i.e. *palmitic acid*. To simplify presentation and discussion of fatty acid compositions, shorthand nomenclatures also exist. In the simplest form, fatty acids are designated solely by the number of carbon atoms they possess, e.g. palmitic acid is a C_{16} acid. This compound can be defined more accurately, however, by listing both the number of carbon atoms in the acid and also the number of double bonds, separating the two figures by a colon, i.e. taking the above example—16:0. Table 1.1 contains a list of common saturated fatty acids together with their trivial and systematic names and shorthand designations.

TABLE 1.1. SATURATED ACIDS OF GENERAL
FORMULA $CH_3.(CH_2)_n.COOH$

Systematic name	Trivial name	Shorthand designation
ethanoic	acetic	2:0
propanoic	propionic	3:0
butanoic	butyric	4:0
pentanoic	valeric	5:0
hexanoic	caproic	6:0
heptanoic	enanthic	7:0
octanoic	caprylic	8:0
nonanoic	pelargonic	9:0
decanoic	capric	10:0
hendecanoic	—	11:0
dodecanoic	lauric	12:0
tridecanoic	—	13:0
tetradecanoic	myristic	14:0
pentadecanoic	—	15:0
hexadecanoic	palmitic	16:0
heptadecanoic	margaric	17:0
octadecanoic	stearic	18:0
nonadecanoic	—	19:0
eicosanoic	arachidic	20:0
heneicosanoic	—	21:0
docosanoic	behenic	22:0
tetracosanoic	lignoceric	24:0

Acetic acid is rarely found in association with higher molecular weight fatty acids in esterified form, although it has been found esterified to glycerol and to hydroxy-fatty acids in some seed oils. C_4 to C_{12} acids are found mainly in milk fats, although the C_{10} and C_{12} acids have also been found in quantity in certain seed oils. *Myristic acid* (14:0) is a minor component of most animal lipids, but is present in major amounts in seed oils of the family Myristicaceae. Palmitic acid is probably the commonest saturated fatty acid and is found in virtually all animal and plant fats and oils. *Stearic acid* (18:0) is also relatively common and may on occasion be more abundant than palmitic acid, especially in complex lipids. Longer chain saturated acids occur less frequently, but are often major components of waxes. C_{15} to C_{19} odd-chain acids can be found in only trace amounts in most animal lipids but can occur in larger quantities in certain species of fish or in bacterial lipids.

Decanoic and higher saturated fatty acids are solids at room temperature. Because of the lack of functional groups other than the carboxyl group, they are comparatively inert chemically and natural or synthetic lipids containing only these acids can be subjected to more vigorous chemical conditions than those containing polyunsaturated fatty acids.

2. *Monoenoic fatty acids*

Straight-chain even-numbered fatty acids of 10 to 30 carbon atoms containing one double bond of the *cis*-configuration have been characterised from natural sources. Monoenoic acids with double bonds in the *trans*-configuration are not unknown but are comparatively rare. Fatty acids of a given chain length may have the double bond in a number of different positions and a full description of any acid must specify the position and configuration of the double bond (in the recommended numbering system, the carboxyl carbon is C-1). For example, by far the commonest monoenoic acid is *cis*-9-octadecenoic acid, less accurately Δ^9-octadecenoic acid, which has the trivial name *oleic acid* and structure

$$CH_3.(CH_2)_7.CH = CH.(CH_2)_7.COOH$$
$$cis$$

In the shorthand nomenclature, the acid is designated 18:1. In addition, the position of the double bond can be denoted in the form $(n\text{-}x)$, where n is the chain length of the acid and x the number of carbon atoms from the last double bond to the terminal methyl group, i.e. oleic acid is $18:1(n\text{-}9)$ or in the older literature $18:1\omega9$. A IUPAC–IUB Commission[235] has reluctantly agreed to the first form of this nomenclature because of its convenience to biochemists with interests in fatty acid metabolism in animals, as will become apparent later. Table 1.2 contains a list of the common monoenoic acids together with their systematic and trivial names and their shorthand designations.

TABLE 1.2. THE MORE IMPORTANT MONOENOIC
ACIDS OF GENERAL FORMULA
$CH_3.(CH_2)_m.CH=CH.(CH_2)_n.COOH$

Systematic name	Trivial name	Shorthand designation
cis-9-dodecenoic	lauroleic	12:1 (*n*-3)
cis-9-tetradecenoic	myristoleic	14:1 (*n*-5)
trans-3-hexadecenoic	—	16:1*
cis-9-hexadecenoic	palmitoleic	16:1 (*n*-7)
cis-6-octadecenoic	petroselinic	18:1 (*n*-12)
cis-9-octadecenoic	oleic	18:1 (*n*-9)
trans-9-octadecenoic	elaidic	18:1*
is-11-octadecenoic	*cis*-vaccenic	18:1 (*n*-7)
trans-11-octadecenoic	*trans*-vaccenic	18:1*
cis-9-eicosenoic	gadoleic	20:1 (*n*-11)
cis-11-eicosenoic	gondoic	20:1 (*n*-9)
cis-13-docosenoic	erucic	22:1 (*n*-9)
cis-15-tetracosenoic	nervonic	24:1 (*n*-9)

* The (*n-x*) nomenclature is only used with fatty acids containing *cis* double bonds.

Oleic acid is probably the most abundant fatty acid of all and is found in virtually all lipids of animal and plant origin. Various positional isomers exist; for example, *petroselinic acid* (*cis*-6-octadecenoic acid) occurs in seed oils of the family Umbelliferae and *cis-vaccenic acid* (*cis*-11-octadecenoic acid) is the major unsaturated acid in many bacterial species. The *trans*-isomer of oleic acid (*elaidic acid*) is rarely found, but *trans*-vaccenic acid, which is a by-product of the biohydrogenation of polyunsaturated fatty acids in the rumen, is found in small amounts in the lipids of ruminant animals. Many different positional isomers of monoenoic fatty acids may be present in a single natural lipid; for example, five different *cis*-octadecenoic acids and eleven different *trans*-octadecenoic acids have been found in bovine milk triglycerides.[195]

Palmitoleic acid is a component of most animal fats and may be present in quite high concentrations in fish oils and some seed oils.

Shorter-chain monoenoic acids are common constituents of milk fats, but are rarely found in significant amounts in other tissues. C_{20} and C_{22} monoenoic acids are minor components of most animal lipids, but are found in appreciable quantities in certain seed oils (e.g. rape seed oil) and in fish oils. Odd-chain monoenoic acids are sometimes found as minor components of animal lipids but may be present in larger amounts in some fish oils and in bacterial lipids.

Animal lipids frequently contain families of monoenoic fatty acids with different chain lengths but with similar terminal structures. Here

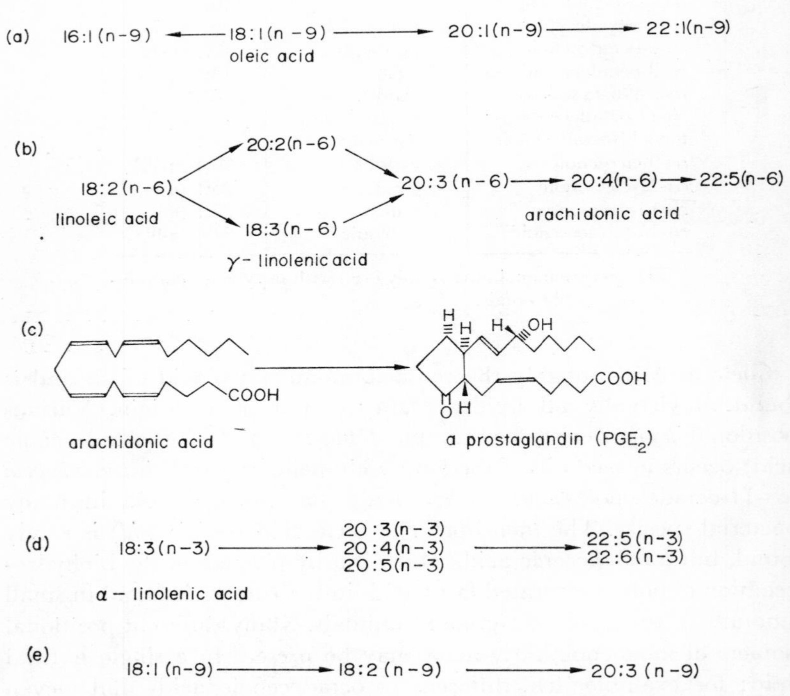

(a) $16:1(n-9)$ ◄——— $18:1(n-9)$ ———► $20:1(n-9)$ ———► $22:1(n-9)$
                        oleic acid

(b)
                    $20:2(n-6)$
          ↗   ↘
$18:2(n-6)$ $20:3(n-6)$ ———► $20:4(n-6)$ ——► $22:5(n-6)$
linoleic acid   ↘   ↗   arachidonic acid
                    $18:3(n-6)$
                  γ- linolenic acid

(c)

arachidonic acid   a prostaglandin (PGE$_2$)

(d) $20:3(n-3)$
       $18:3(n-3)$ ———————► $20:4(n-3)$ ——————— $22:5(n-3)$
                        $20:5(n-3)$ $22:6(n-3)$
    α - linolenic acid

(e)
       $18:1(n-9)$ ———————► $18:2(n-9)$ ———————► $20:3(n-9)$

Fig. 1.1. Biosynthetic relationships between unsaturated fatty acids. a, Elongation and retroconversion of oleic acid. b, Elongation and desaturation of linoleic acid. c, Biosynthesis of prostaglandin E$_2$ from arachidonic acid. d, Elongation and desaturation of α-linolenic acid. e, Elongation and desaturation of oleic acid.

biosynthetic relationships are obvious as several components may arise by chain elongation or chain shortening of a common precursor as illustrated in Fig. 1.1a.

cis-Monoethenoid fatty acids of eighteen carbon atoms or less are all low melting compounds. *trans*-Isomers generally have slightly higher melting points than the corresponding *cis*-compounds. Because of the presence of the double bond in the aliphatic chain, monoenoic acids and lipids containing such acids are more susceptible to chemical attack, in particular by oxidising agents, than the corresponding saturated compounds. They are fairly resistant to autoxidation but will succumb to this under very vigorous conditions.

3. Non-conjugated polyunsaturated fatty acids

Non-conjugated polyunsaturated fatty acids (often abbreviated to PUFA) of animal and plant origin can be subdivided into several simple families according to their biosynthetic derivation from single specific fatty acid precursors.[341] The acids in each family contain two or more *cis*-double bonds separated by a single methylene group (methylene interrupted unsaturation) and have a common terminal structure. Table 1.3 contains a list of the more important of these acids.

TABLE 1.3. THE MORE IMPORTANT NON-CONJUGATED POLYUNSATURATED FATTY ACIDS OF GENERAL FORMULA
$$CH_3 \cdot (CH_2)_m \cdot (CH= CH.CH_2)_x \cdot (CH_2)_n \cdot COOH$$

Systematic name	Trivial name	Shorthand designation
9,12-octadecadienoic*	linoleic	18:2 (*n*-6)
6,9,12-octadecatrienoic	γ-linolenic	18:3 (*n*-6)
8,11,14-eicosatrienoic	homo-γ-linolenic	20:3 (*n*-6)
5,8,11,14-eicosatetraenoic	arachidonic	20:4 (*n*-6)
4,7,10,13,16-docosapentaenoic	—	20:5 (*n*-6)
9,12,15-octadecatrienoic	α-linolenic	18:3 (*n*-3)
5,8,11,14,17-eicosapentaenoic	—	20:5 (*n*-3)
4,7,10,13,16,19-docosahexaenoic	—	22:6 (*n*-3)
5,8,11-eicosatrienoic	—	20:3 (*n*-9)

* The double-bond configuration in each instance is *cis*.

Linoleic acid (*cis*-9, *cis*-12-octadecadienoic acid) is the commonest and simplest fatty acid of this type and is found in most plant and animal tissues. Using the same shorthand nomenclature as before, it can be designated 18:2 (*n*-6) (it is assumed that there is a methylene interrupted double bond system). It is an *essential* fatty acid (often abbreviated to EFA) in animal diets as it cannot be synthesised by the animal yet is required for growth, reproduction and healthy development.[215] In animal tissues, it is the precursor of a family of other fatty acids which are produced from it by desaturation and chain elongation as illustrated in Fig. 1.1b. All have the (*n*-6) terminal structure and can also function as essential fatty acids. The enzymes in mammalian systems are only able to insert double bonds between the carboxyl group and the first double bond already present in the fatty acid while plant enzyme systems can only insert new double bonds between the last double bond and the terminal portion of the fatty acid.

Arachidonic acid (*cis*-5, *cis*-8, *cis*-11, *cis*-14-eicosatetraenoic acid) is the most important of the linoleic acid metabolites and is a major constituent of the complex lipids of animal tissues but is rarely found in plants. It is one of the principal precursors of a highly important series of hormone-like compounds known as prostaglandins[420] as shown in Fig. 1.1c. These compounds have profound pharmacological activity and are currently the subject of intensive study; it now appears possible that there is a connection between certain of the symptoms of EFA deficiency and the presence or absence of prostaglandins.

γ-Linolenic acid (*cis*-6, *cis*-9, *cis*-12-octadecatrienoic acid), an important intermediate in the biosynthesis of arachidonic acid, occurs in only minor amounts in animal tissues but is found in appreciable quantities in some seed oils.

Linolenic acid (sometimes termed *α*-linolenic acid) or *cis*-9, *cis*-12, *cis*-15-octadecatrienoic acid (18:3 (*n*-3)) is a major component of plant lipids, particularly of the photosynthetic tissues, but is rarely a significant constituent of animal lipids. It is none the less extremely important as the primary precursor of another important family of polyunsaturated fatty acids (Fig. 1.1d). Linolenic acid and/or polyunsaturated fatty acids of the (*n*-3) family are essential fatty acids in fish,[487] but the weight of evidence tends to indicate that they are not essential in mammals. 5,8,11,14,17-Eicosapentaenoic acid and

4,7,10,13,16,19-docosahexaenoic acid, in particular, are found in many animal tissues as major components of the complex lipids and they are also found in large amounts in fish oils.

Oleic acid can also be the primary precursor of a family of poly-unsaturated fatty acids (Fig. 1.1e). 5,8,11-Eicosatrienoic acid is normally a minor component of animal lipids but can assume signi-ficance in the complex lipids of animals deficient in essential fatty acids. There is also a family of polyunsaturated fatty acids derived from palmitoleic acid (n-7), and odd-chain polyunsaturated fatty acids have been found in some fish oils.

Polyunsaturated acids with more than one methylene group between the double bonds have been found in some bacterial and plant lipids but are rarely found in animals. Isomers of linoleic acid in which one or more double bonds have a *trans*-configuration have also been found in seed oils.

Polyunsaturated fatty acids all have very low melting points. The greater the number of double bonds they possess, the greater their susceptibility to autoxidation. If the acids or their derivatives are subjected to too high temperatures or to alkaline hydrolysis under conditions that are too vigorous, migration or stereomutation of double bonds can occur.

4. *Branched-chain and cyclopropane fatty acids*

Branched-chain fatty acids are common constituents of bacterial lipids.[2,399] Those found most frequently have a single methyl group on the penultimate (*iso*) or antepenultimate (*anteiso*) carbon atoms (Fig. 1.2), although simple methyl-branched acids with the substituent elsewhere in the chain are occasionally found (e.g. D-(−)-10-methyl-stearic acid or "tuberculostearic acid"). Shorter-chain multibranched fatty acids are major constituents of avian preen glands. *Phytanic acid* (3,7,11,15-tetramethyl-hexadecanoic acid, Fig. 1.2), a metabolite of phytol, is present in trace amounts in animal adipose tissue lipids but assumes major proportions in Refsum's syndrome, a rare condition in which a deficiency in the fatty acid *a*-oxidation enzyme system prevents the catabolism of the acid causing it to accumulate. Very high mole-

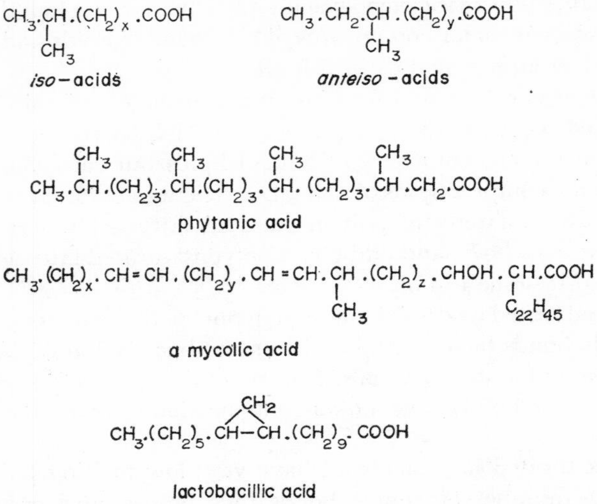

$$CH_3.CH.(CH_2)_x.COOH$$
$$\qquad | $$
$$\qquad CH_3$$

iso – acids

$$CH_3.CH_2.CH.(CH_2)_y.COOH$$
$$\qquad\qquad | $$
$$\qquad\qquad CH_3$$

anteiso – acids

$$\qquad CH_3 \quad\;\; CH_3 \quad\;\; CH_3 \quad\;\; CH_3$$
$$\qquad\; | \qquad\; | \qquad\; | \qquad\; |$$
$$CH_3.CH.(CH_2)_3.CH.(CH_2)_3.CH.(CH_2)_3.CH.CH_2.COOH$$

phytanic acid

$$CH_3.(CH_2)_x.CH=CH.(CH_2)_y.CH=CH.CH.(CH_2)_z.CHOH.CH.COOH$$
$$\qquad\qquad\qquad\qquad\qquad\qquad\qquad | \qquad\qquad\qquad\quad |$$
$$\qquad\qquad\qquad\qquad\qquad\qquad\qquad CH_3 \qquad\qquad\qquad C_{22}H_{45}$$

a mycolic acid

$$\qquad\qquad\qquad CH_2$$
$$\qquad\qquad\qquad / \;\; \backslash$$
$$CH_3.(CH_2)_5.CH-CH.(CH_2)_9.COOH$$

lactobacillic acid

FIG. 1.2. The structures of some branched-chain and cyclic fatty acids.

cular weight branched-chain fatty acids having up to eighty or more carbon atoms, the *mycolic* and related acids (Fig. 1.2), have been found in the lipids of certain bacteria.[399] Branched-chain acids have been found in only one plant species (*Antirrhinum majus*). They are often found in small amounts in the depot fats of ruminant animals, though such acids are probably of bacterial origin in this instance. Sebum and animal waxes, e.g. wool-wax, may contain significant amounts of these acids, however, and isovaleric acid is a major component of dolphin triglycerides. As they are generally fully saturated compounds, branched-chain acids are comparatively resistant to chemical degradation. None the less, their structures can readily be determined by physical means such as mass spectrometry.

Cyclopropane fatty acids,[89] for example *lactobacillic acid* (11,12-methyleneoctadecanoic acid, Fig. 1.2), are found in bacterial lipids, particularly in those of a number of gram-negative and a few gram-positive families of the order Eubacteriales. They are also found, generally in small amounts, in certain seed oils where they may be biosynthetic precursors of cyclopropene fatty acids. In such compounds the cyclopropane ring has some of the properties of a double bond although it may also often behave as a saturated entity; for example, while it is resistant to most oxidising agents, it may be disrupted by strong acids or by halogens.

5. *Unusual fatty acids of plant origin*

In addition to the more common saturated, monoenoic and C_{18} poly-unsaturated fatty acids, plant lipids may contain a wide variety of unusual fatty acids not found in the animal kingdom. The structures of these have been comprehensively reviewed by Smith.[501] So many have been found, indeed, that it is not possible to discuss them all in a brief space, but a list of some of the more important functional groups found in the fatty acid chain will show the potential complexity of the problem of analysing lipids containing such acids. They include acetylenic bonds, conjugated acetylenic and ethylenic bonds, allenic groups, cyclopropane, cyclopropene, cyclopentene and furan rings, epoxyl, hydroxyl and keto groups and double bonds of both the *cis*- and *trans*-configurations and separated by more than one methylene group. Two or more of these functional groups may on occasion be found in a single fatty acid. A brief list of some of the more important of the acids is contained in Table 1.4. Certain of the functional groups may be destroyed by chemical techniques that are used widely in the analysis of animal lipids. For example, epoxyl and cyclopropene rings are disrupted by acidic conditions such as may be used to transesterify lipids.

Most of these acids occur in obscure seed oils, but a few have commercial importance or are important as biosynthetic precursors of other acids. *Ricinoleic acid* (D-(+)-12-hydroxy-octadec-*cis*-9-enoic acid) is by far the major component of castor oil and can be pyrolysed to give a number of products of industrial importance. *Crepenynic acid* (octadec-*cis*-9-en-12-ynoic acid) is important as a precursor of a complex series of poly-ynoic metabolites. The commercial value of cotton seed oil is reduced as it contains cyclopropene fatty acids in amounts which are small but sufficient to produce profound pharmacological effects in

TABLE 1.4. SOME UNUSUAL FATTY ACIDS FOUND IN PLANTS

Structure	Trivial name	Source
$CH_3 \cdot (CH_2)_5 \cdot CH.CH_2.CH=CH.(CH_2)_7.COOH$ $\quad\quad\quad\quad OH \quad\quad cis$	ricinoleic acid	castor oil
$CH_3 \cdot (CH_2)_4.C \equiv C.CH_2.CH=CH.(CH_2)_7.COOH$ $\quad\quad\quad\quad\quad\quad\quad\quad cis$	crepenynic acid	*Crepis foetida*
CH_2 $CH_3 \cdot (CH_2)_7.C = C.(CH_2)_7.COOH$	sterculic acid	$\left\{\begin{array}{l}\text{Sterculiaceae}\\\text{Malvaceae}\end{array}\right.$
$CH_3 \cdot (CH_2)_3.CH=CH.CH=CH.CH=CH.(CH_2)_7.COOH$ $\quad\quad trans \quad trans \quad cis$	α-eleostearic acid	tung oil
$CH_3 \cdot (CH_2)_4.CH{-}CH.CH_2CH=CH(CH_2)_7COOH$ $\quad\quad\quad\quad\quad\searrow O \swarrow \quad\quad cis$ $\quad cis$	(+)-vernolic acid	*Vernonia anthelmintica*
$(CH_2)_{12}COOH$	chaulmoogric acid	*Hydnocarpus* species
$CH_3 \cdot (CH_2)_7.CH=CH.(CH_2)_5.CH=CH.(CH_2)_3.COOH$ $\quad cis \quad\quad\quad\quad\quad\quad cis$	—	*Limnanthes douglasii* seed oil
$CH_3 \cdot (CH_2)_{10}.CH=C=C.(CH_2)_3.COOH$	laballenic acid	*Leonotis nepetaefolia* seed oil

animals by inhibiting fatty acid desaturation when the oil is incorporated into animal feedstuffs.[89]

When the presence of such unusual acids in plant tissues is suspected, therefore, the intact lipids should be examined as fully as possible by non-destructive spectroscopic techniques such as infrared, ultraviolet and nuclear magnetic resonance spectroscopy to determine which functional groups are present in the fatty acid constituents, and indeed such techniques may on occasion be useful in estimating particular components. Deviations from the normal behaviour of standard lipid components on chromatographic adsorbents may also be a useful diagnostic guide in indicating the presence of polar functional groups. When the presence of such groups has been confirmed, appropriate chemical degradative or hydrolytic techniques can then be applied to minimise the destruction of essential structural features in the fatty acid components.

C. Simple Lipids

The simple lipids contain only fatty acid and alcohol components. The alcohol is usually glycerol but may also be a long-chain alcohol or a sterol. Esters of C_2, C_3 and C_4 diols are also known, but are rarely found in greater than trace amounts.[44] Unesterified or free fatty acids, which occur in small amounts in most animal and plant tissues, and free sterols, the most abundant of which is cholesterol, are also classed as simple lipids.

1. *Triglycerides and partial glycerides*

In triglycerides, each hydroxyl group of the trihydric alcohol glycerol is esterified to a fatty acid. If the two primary positions contain different fatty acids, a centre of asymmetry is created and the triglycerides may exist in different enantiomorphic forms so a system of nomenclature must be adopted which takes account of this fact. Several such nomenclatures based on the conventional D/L and R/S systems have been proposed, but a "stereospecific numbering system" is now favoured.[235] In a Fischer projection (Fig. 1.3), the secondary hydroxyl group is

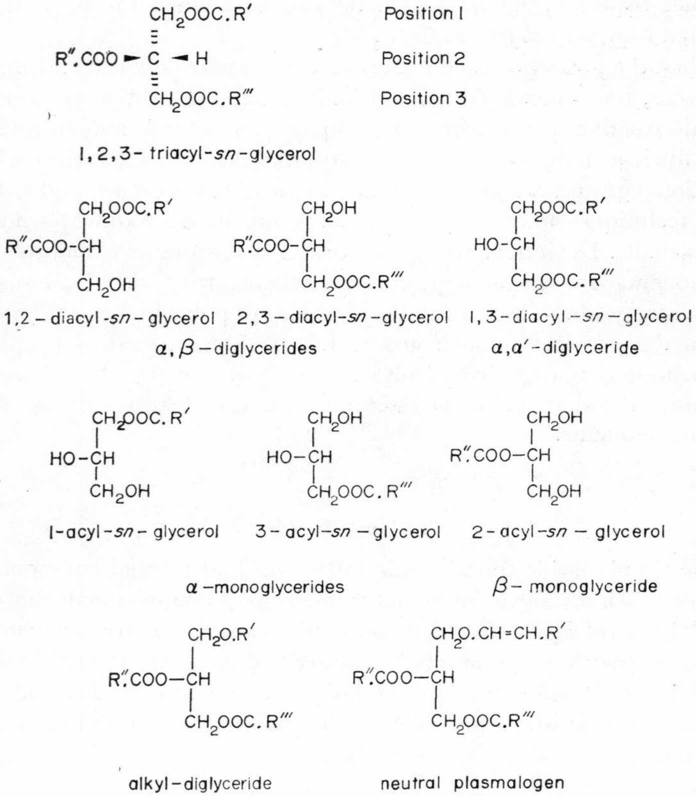

$$CH_2OOC.R' \qquad \text{Position I}$$
$$R''.COO \blacktriangleright C \blacktriangleleft H \qquad \text{Position 2}$$
$$CH_2OOC.R''' \qquad \text{Position 3}$$

I, 2, 3 – triacyl-*sn*-glycerol

$$CH_2OOC.R' \qquad\qquad CH_2OH \qquad\qquad CH_2OOC.R'$$
$$R''.COO-CH \qquad\qquad R''.COO-CH \qquad\qquad HO-CH$$
$$CH_2OH \qquad\qquad CH_2OOC.R''' \qquad\qquad CH_2OOC.R'''$$

1,2 – diacyl-*sn*-glycerol   2,3 – diacyl-*sn*-glycerol   I,3-diacyl-*sn*-glycerol

α, β – diglycerides α, α' – diglyceride

$$CH_2OOC.R' \qquad\qquad CH_2OH \qquad\qquad CH_2OH$$
$$HO-CH \qquad\qquad HO-CH \qquad\qquad R''.COO-CH$$
$$CH_2OH \qquad\qquad CH_2OOC.R''' \qquad\qquad CH_2OH$$

I-acyl-*sn*-glycerol   3-acyl-*sn*-glycerol   2-acyl-*sn*-glycerol

α – monoglycerides β – monoglyceride

$$CH_2O.R' \qquad\qquad CH_2O.CH=CH.R'$$
$$R''.COO-CH \qquad\qquad R''.COO-CH$$
$$CH_2OOC.R''' \qquad\qquad CH_2OOC.R'''$$

alkyl – diglyceride   neutral plasmalogen

FIG. 1.3. The structures of simple glycerides.

shown to the left of C-2; the carbon atom above this then becomes C-1 and that below C-3 and the prefix *sn* is placed before the stem name of the compound. When the stereochemistry of the molecule is not specified, the primary hydroxyl groups are often termed the α- and α'-positions and the secondary the β-position. The topic of glyceride chirality has been discussed in detail by Smith.[503] Triglycerides are quantitatively by far the most important single lipid class and virtually all the commercially significant fats and oils of animal or plant origin and most animal depot fats consist almost entirely of this lipid.

Certain fungal and seed lipids have been found with triglyceride components that contain hydroxy-fatty acids, the hydroxyl group of which is esterified to an additional fatty acid. Such lipids have been termed *estolides*.

Diglycerides and *monoglycerides* contain two moles or one mole of fatty acid per mole of glycerol respectively (Fig. 1.3). They are rarely present in more than trace amounts in fresh animal and plant tissues but the 1,2-diacyl-*sn*-glycerols have particular importance biosynthetically as precursors of triglycerides and complex lipids. The IUPAC–IUB Commission on Biochemical Nomenclature has recommended that the names triacyl, diacyl and monoacyl glycerol should replace the terms triglyceride, diglyceride and monoglyceride respectively as the former defined the structures of the compounds more precisely.

A full analysis of triglycerides and partial glycerides requires that not only the total fatty acid composition of the component be determined but also the distribution of fatty acids in each position. Partial glycerides undergo acyl migration very readily in alcoholic solvents, on heating, or in acidic or basic media and even when standing for long periods in the crystalline state so separatory and analytical procedures should be such as to minimise this occurrence if a knowledge of their structures is required.

2. *Alkyl diglycerides and neutral plasmalogens*

Alkyl diglycerides[505] are lipid components in which a long-chain alkyl group is joined by an ether linkage to position 1 of L-glycerol; positions 2 and 3 are esterified with conventional fatty acids (Fig. 1.3). The alkyl groups of these 1-alkyl-2,3-diacyl-*sn*-glycerols are generally saturated or *cis*-monoenoic even-numbered components with sixteen to twenty carbon atoms. On hydrolysis, fatty acids (2 moles) and 1-alkyl glycerols are obtained. The trivial names "chimyl", "batyl" and "selachyl" alcohol are used for 1-hexadecyl-, 1-octadecyl- and 1-octadec-9′-enyl-glycerol respectively. Alkyl ethers occur in small amounts in many animal tissues and are occasionally found in large quantities in the lipids of marine animals. The ether linkage is stable to acidic or basic conditions although the ester bonds are readily hydrolysed as in more conventional lipids. The alkyl moieties are usually analysed in the form of the 1-alkyl glycerol or as volatile non-polar derivatives of this compound.

Neutral plasmalogens are related compounds in which position 1 of L-glycerol is linked by a vinyl ether bond (the double bond is of the *cis*-configuration) to an alkyl group (Fig. 1.3). These 1-alk-1'-en-1'-yl-2,3-diacyl-*sn*-glycerols are rarely found in other than very small amounts in animal tissues, although heart muscle and nervous tissues may contain appreciable amounts, especially in ruminant animals. Although the vinyl ether linkage is unaffected by basic hydrolysis conditions, it is disrupted by strong acids yielding one mole of an aldehyde.

$$\text{R.O.CH}=\text{CH.R}' \xrightarrow{\quad H_3O^+ \quad} \text{ROH} + \text{OHC.CH}_2.\text{R}'$$

The principal aldehydes usually have sixteen or eighteen carbon atoms and are fully saturated or contain one double bond.

3. *Cholesterol and cholesteryl esters*

Cholesterol is by far the commonest member of a group of steroids with a tetracyclic ring system and in animal tissues it occurs in the free state in intimate association with esterified lipids in cell membranes and in serum lipoproteins. It is also found esterified to fatty acids (*cholesteryl esters*) in animal tissues, especially in the liver, adrenals and plasma. Cholesteryl esters are hydrolysed or transesterified much more slowly than most other O-acyl lipids.

Plant tissues contain related sterols, e.g. β-sitosterol, ergosterol and stigmasterol, but only trace amounts of cholesterol. Steroidal hormones also occur in very small quantities in all animal tissues, but these are not lipids if we apply the definition stated earlier. Similarly, bile acids, which are important metabolites of cholesterol, cannot be discussed in detail here.

4. *Wax esters and other minor components*

Wax esters consist of fatty acids esterified to long-chain alcohols. The fatty acids are usually straight-chain saturated or monoenoic compounds containing up to thirty carbon atoms, but branched-chain and α- or ω-hydroxy acids are also found on occasion. Saturated long-chain primary alcohols usually predominate, but monoenoic, branched-chain and secondary alcohols are sometimes present also. Wax esters are found in animal and insect secretions (e.g. wool wax and beeswax) and in sebum lipids. They also occur as protective coatings on plant leaves and fruits[127] and in marine animals they are sometimes major components of the depot fats. The alcohols are found in the free state only rarely and then in trace amounts.

Long-chain hydrocarbons of similar structure to the wax alcohols are sometimes found in association with wax esters. The triterpenoic hydrocarbon, *squalene*, is a major component of some fish lipids, but is found in only trace quantities in animal or plant tissues. It is important, none the less, as a biosynthetic precursor of cholesterol.

C_{16} and C_{18} saturated or monoenoic straight-chain aldehydes, which may be either precursors or degradation products of plasmalogens, are found in very small amounts in some animal tissues.

D. Complex Lipids

The complex lipids can be subdivided into three main classes: *phosphoglycerides*, which on hydrolysis yield glycerol, fatty acids, inorganic phosphate and an organic base or polyhydroxy compound; *glycosyl diglycerides*, which on hydrolysis yield glycerol, fatty acids and sugars; *sphingolipids*, which contain a long-chain base, fatty acids and inorganic phosphate, carbohydrates or other complex organic compounds. In phosphoglycerides and glycosyl diglycerides, the non-fatty acid components are linked to position 3 of L-glycerol and the stereospecific numbering system described earlier for simple glycerol derivatives is applied. The term "glycolipid" is used to describe any lipid class which contains a sugar residue so includes glycosyl diglycerides and certain of the sphingolipids. The term "phospholipid" denotes any lipid which yields phosphoric acid as a hydrolysis product so includes phosphoglycerides and sphingomyelin.

Phosphoglycerides (or "glycerophosphatides") are found in all plant and animal tissues and in microorganisms; glycerol-containing glycolipids are almost entirely plant and bacterial lipids, although trace amounts have been found in brain and sphingolipids are important animal lipid constituents although they have also been found in plants and in some microorganisms. Although they are soluble in most polar organic solvents, phosphoglycerides are generally insoluble in acetone and this property is often used to separate them from glycolipids and simple lipids.

1. *Phosphoglycerides*

(i) *Phosphatidic acid*

Phosphatidic acid or 1,2-diacyl-*sn*-glycerol-3-phosphate (Fig. 1.4) occurs in very small amounts in animal and plant tissues, yet it is extremely important biosynthetically as it is the precursor of all other phosphoglycerides and of triglycerides. The fatty acid residues can be removed by mild hydrolytic procedures leaving *sn*-glycerol-3-phosphate (L-α-glycerophosphate) which can be further hydrolysed under more vigorous acidic conditions to glycerol and phosphoric acid. *sn*-Glycerol-3-phosphate is indeed the biosynthetic precursor of phosphatidic acid although dihydroxyacetone phosphate may be important in some circumstances. Phosphatidic acid is strongly acidic and is often isolated from tissues in the form of a mixed salt so that it may be necessary to prepare a single salt form to facilitate its purification.

In common with most phosphoglycerides in animal tissues, position 1 tends to be occupied by saturated fatty acids, although some monoenoic fatty acids are also usually present, while position 2 is occupied by polyunsaturated fatty acids.

(ii) *Phosphatidyl glycerol and polyglycerophosphatides*

Phosphatidyl glycerol or 1,2-diacyl-*sn*-glycero-3-phosphoryl-1'-*sn*-glycerol (Fig. 1.4) is present in small quantities in many animal tissues,

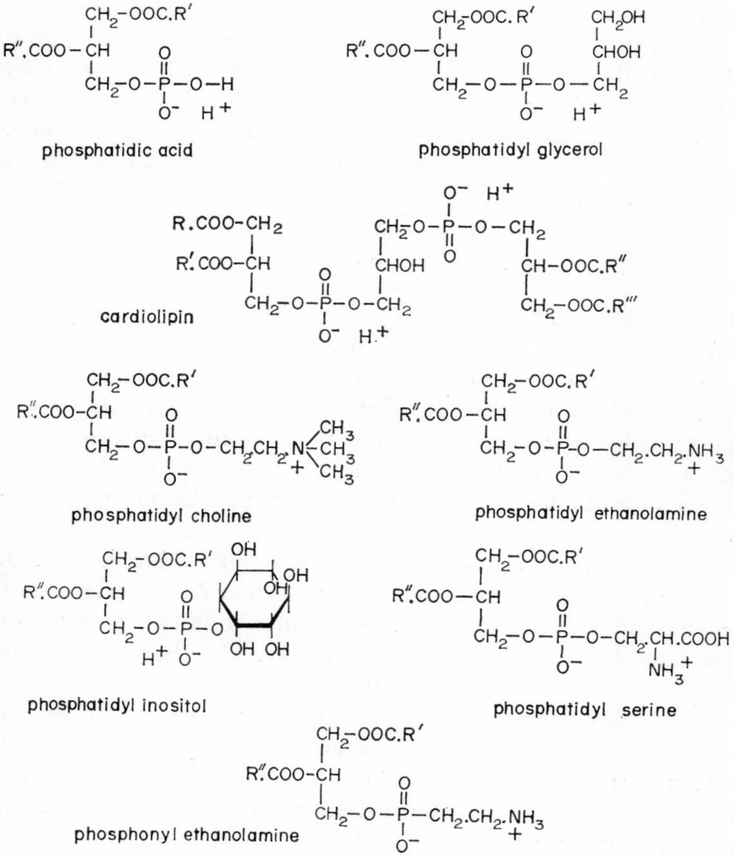

Fig. 1.4. The structures of the principal phosphoglycerides.

especially in cell mitochondria, and is also found in plant chloroplasts. Polyglycerophosphatides, in particular cardiolipin (or diphosphatidyl glycerol), are major components of the lipids of mitochondria, especially in heart muscle. Cardiolipins from some animal species contain very high proportions of linoleic acid and indeed molecular species containing 4 moles of this acid have been isolated. Phosphatidyl glycerol

B

and related compounds are acidic lipids and on hydrolysis yield only glycerol, fatty acids and phosphoric acid in various molar proportions.

Bacterial lipids may contain *lipoamino acids*,[324] O-amino acid esters of phosphatidyl glycerol, in which position 3' of the non-acylated glycerol moiety is linked to an amino acid. Usually only one amino acid (commonly lysine, ornithine or alanine) derivative is found in each bacterial species.

(iii) *Phosphatidyl choline*

Phosphatidyl choline (Fig. 1.4), commonly termed lecithin, is more often than not the most abundant phosphoglyceride in animal tissues and is often a major lipid component of plant tissues and of micro-organisms. Position 1 in phosphatidyl cholines of animal origin is almost invariably occupied largely by saturated fatty acids while position 2 contains all the C_{18}, C_{20} and C_{22} polyunsaturated fatty acids. It is a non-acidic lipid and on hydrolysis yields choline together with the normal phosphoglyceride components.

Small amounts of related compounds with alkyl ether or vinyl ether residues in position 1 and fatty acids in position 2 (the latter are often termed phosphatidal cholines) are sometimes found along with diacyl phosphatidyl choline in some animal tissues. They have recently been detected in trace amounts in plants.

Lysophosphatidyl cholines, in which only one of the two available positions of glycerol is esterified to a fatty acid, are often found in small amounts in tissues when phosphatidyl choline is present. It is generally believed that position 1 is esterified in most instances, but, as the acyl group migrates very readily, this has not been easy to confirm and there is some evidence that both possible isomers may exist naturally in the same tissue.

(iv) *Phosphatidyl ethanolamine*

Phosphatidyl ethanolamine (Fig. 1.4), once given the trivial name cephalin, is the second most abundant class of phosphoglycerides and is present in large quantities in animal and plant tissues and is frequently the major lipid class in bacteria. Phosphatidyl ethanolamines of animal origin usually contain more polyunsaturated fatty acids than do the phosphatidyl cholines from the same tissue and these acids are concentrated in position 2. It is a non-acidic lipid and on hydrolysis yields ethanolamine (1 mol), glycerol (1 mol), fatty acids (2 mol) and phosphoric acid (1 mol).

Alkyl and alkenyl analogues (with the anomalous group in position 1) are occasionally found in the same tissue with the diacyl component.

Lysophosphatidyl ethanolamine, in which only one of the two possible positions in the glycerol moiety is esterified, is found infrequently in some animal and plant tissues. Again, because of the ease with which the acyl group migrates, it is not easy to confirm which position is esterified in the natural state.

N-Methyl and NN-dimethyl derivatives of phosphatidyl ethanolamine are common constituents of the lipids of microorganisms. N-Acyl-phosphatidyl ethanolamines are occasionally found as minor components of plant tissues and they have also been detected in bovine brain.

(v) *Phosphatidyl inositides*

Phosphatidyl inositol (Fig. 1.4) is a common constituent of animal, plant and bacterial lipids. All compounds of this type are derivatives of the optically inactive form of inositol, myoinositol. Position 1 of phosphatidyl inositol of animal origin usually contains a high proportion of stearic acid and position 2 a high proportion of arachidonic acid.

Phosphatidyl inositol is accompanied by large amounts of di- and triphosphoinositides in brain lipids. In the former, position 4' of the myoinositol residue is linked to phosphoric acid and in the latter, positions 4' and 5' are linked to phosphoric acid. More complex polyphosphoinositides exist. The chemistry and biochemistry of these compounds have been reviewed.[194] Phosphoinositides are strongly acidic and are usually isolated in association with cations, such as those of magnesium or calcium.

(vi) *Phosphatidyl serine*

Phosphatidyl serine (Fig. 1.4) is a major component of brain and erythrocyte lipids and is found in small quantities in most animal and plant tissues and in bacteria. It is a weakly acidic lipid and on hydrolysis yields L-serine together with the normal phosphoglyceride components. They are usually isolated as the potassium salts but may also contain calcium, sodium or magnesium ions. N-Acyl phosphatidyl serine has been found in sheep erythrocyte lipids.

(vii) *Phosphonolipids*

Phosphoglycerides which contain phosphonic acid esterified to glycerol and a carbon–phosphorus bond to the nitrogenous base have been found in marine invertebrates and in protozoa.[275] The commonest of these phosphonolipids is that related to phosphatidyl ethanolamine, sometimes termed phosphonyl ethanolamine. The phosphorus–carbon bond is extremely resistant to acid hydrolysis and the main products of this reaction are glycerol (1 mol), fatty acids (2 mols) and 2-aminoethylphosphonic acid (1 mol).

Phosphonolipid analogues of phosphatidyl choline, N-methyl- and NN-dimethyl-phosphatidyl ethanolamine and phosphatidyl serine have also been found in addition to the ceramide derivatives discussed below.

2. *Glycosyl diglycerides and related compounds*

(i) *Plant glycosyl diglycerides*

Plant tissues contain lipids in which 1,2-diacyl-*sn*-glycerols are joined by a glycosidic linkage through position 3 to sugar moieties.[205,264] The principal components are mono- and digalactosyl diglycerides (as illustrated in Fig. 1.5), the hydrolysis products of which are fatty acids (2 mols), glycerol (1 mol) and galactose (1 or 2 mols respectively). Glycosyl diglycerides are often present in trace amounts in animal tissues also where they are frequently overlooked. 6-O-Acyl-monogalactosyl diglycerides are found in plants[197] and although they were originally thought to be artefacts, they are now thought to be normal constituents of the tissue.

Plant tissues also contain a unique lipid, the *plant sulpholipid* or monogalactosyl diglycerides.

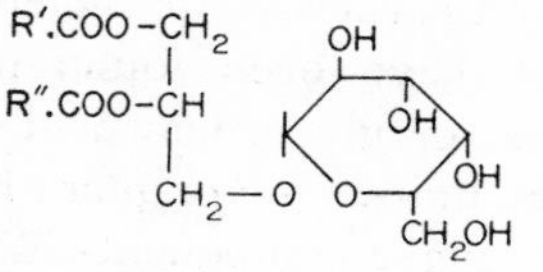

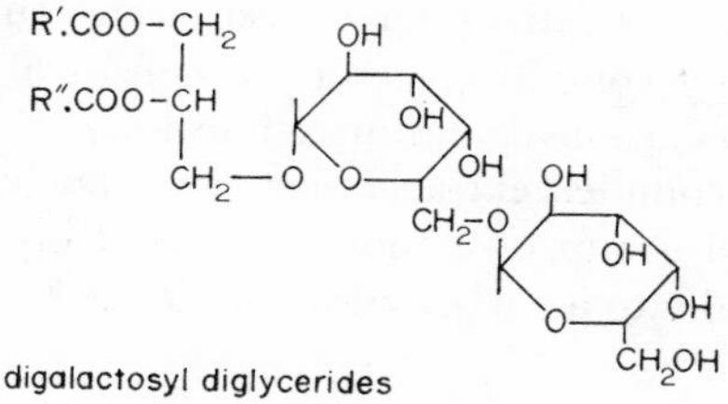

digalactosyl diglycerides

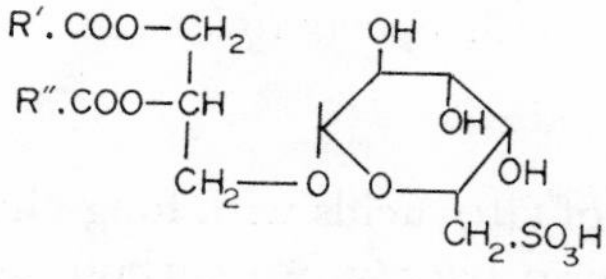

sulphoquinovosyl diglyceride

FIG. 1.5. The structures of glycosyl diglycerides.

sulphoquinovosyl diglyceride, that is found exclusively in the chloro-plasts. This consists of monogalactosyl diglyceride in which position 6 of the galactose moiety is linked by a carbon–sulphur bond to sulphonic acid. Glycolipids are soluble in acetone and are frequently separated from phospholipids by chromatographic systems in which this solvent is used.

A steryl glycoside, 6-O-acyl-β-D-glucopyranosyl-(1-3')-β-sitosterol, has also been found in plant tissues (the sterol component may vary).

(ii) *Bacterial glycolipids*

A large number of glycosyl diglycerides have been isolated from bacterial species. These are 1,2-diacyl-*sn*-glycerols in which position 3

is linked by an ether bond to a carbohydrate moiety. Mono-, di-, tri-
and tetraglycosyl derivatives have been found in which glucose,
galactose, mannose, rhamnose or glucuronic acid in various molar
proportions may be the sugar units. One or more fatty acid residues
may also be esterified to the sugar components. The structures and
occurrence of this complex group of lipids have recently been reviewed
in greater detail than is possible here.[480]

Acylated sugar derivatives have also been found in bacteria.[480]
These lipids do not contain glycerol but consist of one or more fatty
acids esterified to carbohydrates or to inositol.

Yeasts secrete complex extracellular lipids known as *sophorosides* in
which a hydroxyl group of a mono-, di- or trihydroxy fatty acid is
linked glycosidically to a carbohydrate moiety.[519]

3. *Sphingolipids*

(i) *Long-chain bases and ceramides*

Ceramides are amides of fatty acids with long-chain di- or trihydroxy
bases containing twelve to twenty-two carbon atoms in the aliphatic
chain of which the commonest is sphingosine (4-sphingenine or D-
erythro-1,3-dihydroxy-2-amino-*trans*-4-octadecene) as shown in Fig.
1.6. More than sixty long-chain bases have been found in animals,
plants and microorganisms[254,255] and indeed more than thirty of these
have been found in a single tissue.[364] The monounsaturated dihydroxy
base is that found most often in animal tissues, but saturated, diun-
saturated and branched-chain dihydroxy bases have also been found.
Saturated and monounsaturated straight-chain and branched-chain
trihydroxy bases also occur, however. For example, the commonest
plant long-chain base is phytosphingosine (4-D-hydroxy-sphinganine
or D-ribo-1, 3, 4-trihydroxy-2-amino-octadecane, Fig. 1.6). Such bases
are frequently designated by a shorthand nomenclature similar to that
used for fatty acids; the chain length and number of double bonds are
denoted in the same manner with the prefix *d* or *t* to designate di- and
trihydroxy bases respectively, e.g. sphingosine is d18:1 and phytosphingosine is t18:0.

The acyl groups of ceramides are long-chain (up to C_{26}) saturated or

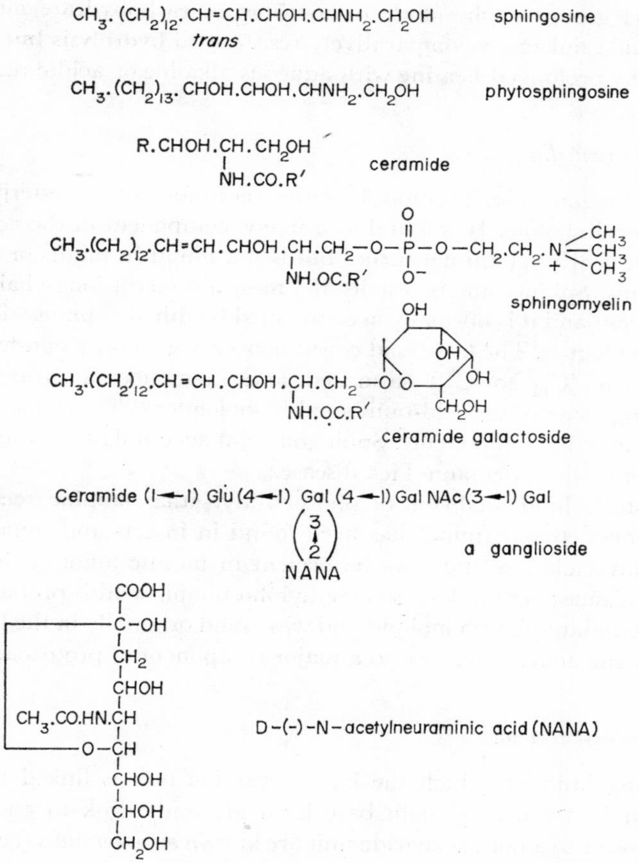

$$CH_3.(CH_2)_{12}.CH{=}CH.CHOH.CHNH_2.CH_2OH \qquad \text{sphingosine}$$
trans

$$CH_3.(CH_2)_{13}.CHOH.CHOH.CHNH_2.CH_2OH \qquad \text{phytosphingosine}$$

$$R.CHOH.CH.CH_2OH$$
$$| \atop NH.CO.R'$$
ceramide

$$CH_3.(CH_2)_{12}.CH{=}CH.CHOH.CH.CH_2{-}O{-}\overset{\overset{O}{\|}}{P}{-}O{-}CH_2.CH_2.N{\overset{+}{<}}{\overset{CH_3}{\underset{CH_3}{-CH_3}}}$$
$$| \atop NH.OC.R' \qquad O^-$$
sphingomyelin

$$CH_3.(CH_2)_{12}.CH{=}CH.CHOH.CH.CH_2{-}O$$
$$| \atop NH.OC.R'$$
ceramide galactoside

Ceramide (1→1) Glu (4→1) Gal (4→1) Gal NAc (3→1) Gal
$$\left(\begin{array}{c}3\\\downarrow\\2\end{array}\right)$$
$$NANA$$
a ganglioside

$$\begin{array}{l}COOH\\|\\C{-}OH\\|\\CH_2\\|\\CHOH\\|\\CH_3.CO.HN.CH\\|\\O{-}CH\\|\\CHOH\\|\\CHOH\\|\\CH_2OH\end{array}$$
D–(–)–N–acetylneuraminic acid (NANA)

FIG. 1.6. The structures of sphingolipids and gangliosides and of their component parts.

monoenoic fatty acids and long-chain 2-D-hydroxy fatty acids. Tetra-
acetyl sphingosines have been isolated from the extracellular lipids of
yeasts.[519] Free ceramides have been found in small amounts in plant
and animal tissues, but generally they form the basic structural unit
of the more complex sphingolipids and are linked through the primary

hydroxyl group to phosphate or to complex carbohydrate moieties. The amide linkage is comparatively resistant to hydrolysis but is disrupted by prolonged heating with aqueous alkaline or acidic reagents.

(ii) *Sphingomyelin*

In sphingomyelin, position 1 of the ceramide unit is esterified to phosphoryl choline. It is found as a major component of the complex lipids in nearly all animal tissues but is not found in plants or microorganisms. Sphingosine is usually the most abundant long-chain base component and it is normally accompanied by dihydrosphingosine and C_{20} homologues. The fatty acid constituents are generally non-hydroxy long-chain (C_{16} to C_{26}) saturated and *cis*-monoenoic compounds including some of the odd-numbered homologues (23:0, in particular, is often a major component). Sphingomyelin accumulates in the tissues of patients with Niemann–Pick disease.

A sphingolipid analogue of phosphatidyl ethanolamine (ceramide phosphoryl ethanolamine) has been found in insects and some freshwater invertebrates and may be present in minute amounts in some animal tissues; ceramide-2-aminoethylphosphonic acid is probably the most abundant phosphonolipid and was found originally in the lipids of the sea anemone,[275] but is also a major component of protozoa.

(iii) *Glycosyl ceramides*

Sphingolipids in which the basic ceramide unit is linked through position 1 of the long-chain base by a glycosidic link to glucose or galactose or to a polysaccharide unit are known as *cerebrosides* (ceramide monohexosides) or *ceramide polyhexosides* respectively. The term "cerebroside" is occasionally used incorrectly to embrace ceramide polyhexosides. The occurrence, chemistry and biochemistry of these compounds have been reviewed.[340,582] As the name suggests, cerebrosides were first found in brain lipids, although they are minor components of most animal tissues and they have also been found in plants (where the hexose is glucose). The major brain cerebroside is a ceramide galactoside (Fig. 1.6). Ceramide glucosides are more common components of other animal tissues, however, particularly in kidney, liver, plasma and spleen. They are also found in plants and in this instance the major long-chain base is most commonly phytosphingosine.

Ceramide di-, tri- and tetrahexosides are also found in animal tissues. The first of these may contain two galactose units (kidney only) or one galactose and one glucose unit (lactosyl ceramide); ceramide trihexosides contain two galactose and one glucose unit; ceramide tetrahexosides (originally given the trivial name "globosides") contain two galactose, one glucose and one galactosamine unit. More complex glycosyl ceramides have been isolated, but as yet have not been fully characterised. The fatty acid residues tend to be principally 2-hydroxy saturated or monoenoic components.

Sulphate esters of cerebroside with the sulphate group linked to position 3 of the galactosyl moiety are major components of brain lipids and are found in trace amounts in other tissues. They are known as *cerebroside sulphates* or *sulphatides* and their chemistry and biochemistry have been reviewed.[177] Ceramide dihexoside sulphate (sulphate derivative of galactosyl-glucosyl ceramide with the sulphate group on position 3 of the galactose unit) has also been isolated and characterised.

Cerebroside and the ceramide polyhexosides have been intensively studied as they have a tendency to accumulate in certain rare disease states such as Gaucher's disease (accumulation of glucose cerebroside), Fabry's disease (ceramide trihexoside) and metachromatic leukodystrophy (cerebroside sulphate).

(iv) *Gangliosides*

Gangliosides are highly complex types of ceramide polyhexosides which contain one or more sialic acid groups (N-acyl derivatives of neuraminic acid, often abbreviated to NANA (see Fig. 1.6).[278,340,582] They are termed mono-, di-, tri- and tetrasialogangliosides according to the number of sialic acid residues which are present in the molecule. In addition, they contain glucose, galactose and N-acetyl galactosamine units. These compounds were given the name "gangliosides" as they are found in high concentration in the ganglion cells of the central nervous system, but they are also found, albeit in much lower concentrations, in other tissues, although the N-acyl group of the sialic acid residue may differ from tissue to tissue or from species to species.

For example, N-glycolyl-neuraminic acid derivatives are often detected in some tissues. The major monosialoganglioside of human brain has the structure shown in Fig. 1.6.

Sphingosine and its C_{20} homologue with smaller amounts of the related dihydro-compounds are the main long-chain bases in brain gangliosides and a very high proportion of the fatty acid components of gangliosides from brain tissue is made up of stearic acid while no hydroxy acids are found. In other tissues, the fatty acid pattern is similar to that of the glycosphingolipids. Gangliosides are soluble both in polar organic solvents and in water so must be extracted from tissues with care. In Tay–Sachs disease, there is a marked increase in the proportions of specific gangliosides in the brain lipids.

(v) *Phytoglycolipid*

This complex lipid, which has been found in a wide variety of plant seeds, consists of ceramide, in which phytosphingosine is the principal long-chain base, linked through a phosphoric acid residue to an oligosaccharide. Inositol, hexuronic acid, hexosamine, mannose, galactose, arabinose and fucose have been identified as components and though a partial structure of the lipid has been obtained, full characterisation has yet to be achieved.[78]

E. Structural Features of Lipids Important in Analyses

The depth of the approach of any research worker to the analysis of lipids will depend on his own specific requirements. For many, for example in simple nutritional experiments, it may only be necessary to isolate and identify the fatty acid components and to determine the overall fatty acid composition of the total lipids. Biochemists, on the other hand, may have more rigorous requirements and may hope to isolate, identify and estimate all the various lipid classes in a tissue and further to estimate the composition of the fatty acid and other components of these lipid classes. His ultimate aim may be to determine the composition of fatty acids esterified in each position of all the glycerol-containing lipids and to isolate and determine single molecular species, which contain only one specific fatty acid in each position, of these lipids.

The first step in any analysis is the isolation of lipids from tissues by extraction with organic solvents and the removal of non-lipid contaminants from these extracts. Procedures have been developed to ensure that there is no loss of the more water-soluble lipids such as gangliosides (see Chapter 2).

If the fatty acid composition of the total lipids is required, the lipids are generally converted to the methyl ester derivatives of the fatty acids by an appropriate procedure (see Chapter 4) for gas chromatographic analysis. With animal lipids, this is usually a straight-forward process, though care must be taken to minimise autoxidation of PUFA and it may be necessary to isolate individual fatty acids at times to determine the position of the double bonds. Plant lipids must be handled with a little more circumspection, however, and it may be necessary to examine these by adsorption chromatography and by spectroscopic procedures to detect unusual functional groups in the fatty acid components which might be destroyed under certain esterification conditions.

Single classes of lipid can be isolated from mixtures by combinations of chromatographic procedures which differentiate lipids according to the number, polarity and acidity or basicity of their constituent parts. They can often be provisionally identified by their chromatographic behaviour relative to that of authentic standards and by spectroscopic techniques and specific spray reagents, but unambiguous identification may require that the various products of hydrolysis be isolated, characterised and estimated.

Enzymatic hydrolysis procedures have been developed for determining the positional distribution of the fatty acid components of a number of simple and complex glycerol-containing lipids (see Chapter 9). Finally, single or at least simpler molecular species of individual lipid classes can be isolated by combinations of procedures which depend on the degree of unsaturation and/or the combined chain lengths of their aliphatic components. Enzymatic hydrolysis may be required to obtain a complete knowledge of the structures of these simpler species.

CHAPTER 2

The Isolation of Lipids from Tissues

QUANTITATIVE isolation of lipids free of non-lipid contaminants must ideally be achieved before the lipid analysis itself can be begun. Carelessness at this preliminary stage may result in the loss of specific components or in the production of artefacts. Lipids can be readily extracted from tissues by a number of organic solvents, but special precautions are necessary to ensure that the recovery is complete. Non-lipid contaminants must then be eliminated from the extract by washing or by column chromatography procedures before the sample is ready for analysis. Precautions must be taken at each stage to minimise the risk of autoxidation of polyunsaturated fatty acids or of hydrolysis of lipids.

A. General Principles of Solvent Extraction Procedures

1. *Storage of tissues*

Ideally, animal, plant or bacterial tissues should be extracted as soon as possible after removal from the parent living organisms so that there is little opportunity for changes to occur in the lipid components. This is especially important with brain lipids. When this is not feasible, the tissue should be frozen as rapidly as possible, for example, on dry ice, and stored in sealed glass containers at $-20°C$ in an atmosphere of nitrogen. The freezing process may permanently damage the tissue as the osmotic shock disrupts the cell membranes so that the original environment of the tissue lipids is altered and lipids may come in contact with enzymes from which they are normally protected. In particular, lipolytic enzymes are released that may hydrolyse the lipids on prolonged standing, even at $-20°C$, or on thawing and

30

contact with organic solvents can aid the process. The presence of large amounts of unesterified acids in animal or plant tissues may be an indication that some irreversible damage to the tissues and subsequently to the lipids has occurred. Also, in plant tissues in particular, phospholipase D may be released to attack phospholipids so that there is an accumulation of phosphatidic acid. The lipases in small samples of plant[193] or animal[152] tissues can be deactivated by plunging the tissues into boiling water for brief periods and the shelf-life of material treated in this way is considerably prolonged. It is occasionally recommended that tissues be stored in saline solution, but Holman[213] has advised that such samples be stored under chloroform in all-glass containers or in bottles with teflon-lined caps at −20°C. Eventually, the tissue samples should be homogenised and extracted with solvent without being allowed to thaw.

2. *The solubility of lipids in organic solvents*

Pure single lipid classes are soluble in a wide variety of organic solvents, but many of these are not suitable for extracting lipids from tissues as they are not sufficiently polar to overcome the strong forces of association between tissue lipids and the other cellular constituents such as proteins. None the less, polar complex lipids, which do not normally dissolve readily in non-polar solvents, may on occasion be extracted by these when they are in the presence of large amounts of simple lipids such as triglycerides so that the precise behaviour of any given solvent as a lipid extractant can rarely be predicted. The ideal solvent or solvent mixture for extracting lipids from tissues should be sufficiently polar to remove all lipids from their association with cell membranes or with lipoproteins but should not react chemically with those lipids. At the same time, it should not be so polar that triglycerides and other non-polar simple lipids do not dissolve and are left adhering to the tissues. The extracting solvent may also have a function in preventing any enzymatic hydrolysis of lipids if chosen carefully, otherwise it may actually stimulate side reactions.

The two main features of lipids which affect their solubility in organic solvents are the non-polar hydrocarbon chains of the fatty acid moieties and any polar functional groups such as phosphate or sugar residues. Lipids which contain no markedly polar groups, for example triglycerides or cholesteryl esters, are highly soluble in hydrocarbon solvents such as hexane, benzene or cyclohexane and also in slightly more polar solvents such as chloroform or diethyl ether; they are rather insoluble, however, in highly polar solvents such as methanol. Their solubility in alcoholic solvents increases with the chain length of the hydrocarbon moiety of the alcohol so they are generally more soluble in ethanol and completely soluble in n-butanol. Similarly, the shorter the chain length of the fatty acid residues, the greater the solubility of the lipid in more polar solvents; tributyrin is completely soluble in methanol, while tripalmitin is virtually insoluble in this solvent. Polar lipids, on the other hand, may be only sparingly soluble in hydrocarbon solvents unless solubilised by association with other lipids, but they dissolve readily in more polar solvents such as methanol, ethanol or chloroform.

It is generally believed that no single pure solvent is suitable as a general purpose lipid extractant, although this is strongly disputed by Lucas and Ridout[315] who have presented evidence that 20 volumes (ml per g of tissue) of ethanol will extract essentially all the lipids from liver homogenates in 5 min under reflux. Unfortunately they do not appear to have extended their work to other tissues. Ethanol–diethyl ether (3:1 or 9:1 v/v) mixtures are used more frequently, in particular to remove lipids from lipoprotein fractions. Diethyl ether or chloroform alone are good solvents for lipids but poor extractants of complex lipids from tissues. On the other hand, they are very useful solvents for removing non-polar lipids from triglyceride-rich tissues such as adipose tissue or oil seeds as they do not co-extract significant amounts of non-lipid contaminants. Unfortunately, these solvents may also promote the action of phospholipase D when used to extract plant tissues.[261] n-and *iso*-Propanol inhibit this reaction and the latter isomer, which is the lower boiling of the two, is frequently used, especially as a preliminary extractant, with plant tissues.

Acetone is a poor solvent for phospholipids and is often used to precipitate such compounds from solution in other solvents. The favoured technique is to take up the lipid mixture in diethyl ether and then to add 4 volumes of cold anhydrous acetone when much of the phospholipids precipitate.[189] This procedure is useful for preparing a phospholipid-rich fraction in bulk, but such preparations are of little use in lipid analysis as the phospholipids obtained do not accurately represent those in the original lipid mixture and the solvent partition system described in Chapter 7 is better for the purpose. Tissue water and mutual solubility effects with other lipid components may permit acetone to extract more phospholipids from animal or plant tissues than might be predicted from a knowledge of the solubility of pure lipids in the dry solvent. Acetone may also react chemically with certain lipids (see below). Glycolipids are soluble in acetone and chromatographic solvents in which this solvent is employed are frequently used in the separation of glycolipids from phospholipids.

It now appears to be generally agreed that a mixture of chloroform and methanol in the ratio of 2:1 (v/v) will extract lipid more exhaustively from animal, plant or bacterial tissues than most other simple solvent systems. The more common simple and complex lipids can be readily extracted with this mixture and indeed the only important exceptions are the polyphosphoinositides of brain where it may be necessary to add acid[575] or inorganic salts[349] to the solvents to obtain quantitative yields. The tissue may be homogenised initially in the presence of both solvents, but better results are often obtained if the methanol (10 volumes; ml/g of tissue) is added first followed, after brief blending, by 20 volumes of chloroform. More than one extraction may be necessary, but with most tissues the lipids are removed almost completely after two or three treatments. There is in general no need to reflux the solvents with the tissue homogenate, but this may on occasion be necessary with wet bacterial cells.[565] However, the extractability of tissue is variable and depends both on the nature of the tissue and of the lipids. For example, the extractability of gangliosides from brain is reduced if the concentration of monovalent cations in the tissue is reduced by dialysis or by washing and if the pH of the tissue is lowered.[512] It may be necessary in some instances to employ more stringent extraction procedures; Ways and Hanahan[570] recommend separate re-extraction with chloroform on its own followed by methanol alone, while others[446] advise that a five-stage extraction procedure using both acidic and basic solvent systems may be more successful.

Water is a poor solvent for all lipids but water–methanol mixtures such as are obtained on washing chloroform–methanol extracts to remove non-lipid contaminants may dissolve significant amounts of the more polar lipids such as gangliosides. Solvent systems containing a little water are sometimes recommended for extracting lipids from very difficult materials. As an example, water-saturated butanol[344] is frequently used to extract lipids from cereals or wheat flour.

3. *Removal of non-lipid contaminants*

Most polar organic solvents used to extract lipids from tissues also extract significant amounts of non-lipid contaminants such as sugars, urea, amino acids and salts. When the polar solvents have been removed by evaporation or distillation, the lipids may be taken up in a small volume of a relatively non-polar solvent such as hexane: choloroform (3:1 v/v)[315] leaving many of the extraneous non-lipid substances behind. Separations of this kind are rarely complete so the procedure is little used nowadays although it should not be discounted where large numbers of samples have to be purified for routine analysis, particularly for some of the simple lipid components. Other procedures which have been tried but with only limited success include dialysis, adsorption and cellulose column chromatography, electrodialysis and electrophoresis.

Most of the contaminating compounds can be removed from chloroform–methanol (2:1 v/v) mixtures simply by shaking the combined solvents with one quarter their total volume of water, or better of a dilute salt solution (e.g. 0·88 per cent potassium chloride solution).[138] The solvents partition into a lower phase of composition, chloroform–methanol–water in the ratio 86:14:1 (v/v/v) and an upper phase in which the proportions are 3:48:47 (v/v/v) respectively. The lower phase, which comprises about 60 per cent of the total volume, contains the purified lipid and the upper phase contains the non-lipid contaminants together with any gangliosides which may have been present (varying amounts of other glycolipids may also be in this layer on occasion[480]). If these are minor components or are not required for further analysis, then a simple washing procedure of this kind yields satisfactory lipid samples. Indeed, gangliosides can be recovered from the upper phase by dialysis and lyophilisation[248] (see also Chapter 7). It is extremely important that the proportions of chloroform–methanol– water in the combined phases should be reasonably close to 8:4:3 (v/v/v). If it is necessary to wash the lower phase again, methanol– water (1:1 v/v), i.e. a mixture similar in composition to the upper phase, should be used to maintain these proportions otherwise losses of polar lipids may be greater than expected. Bligh and Dyer[52] have developed a related procedure which is more suitable for large samples as it uses smaller volumes of chloroform and methanol and the water already present in the samples is taken into account when adding further water in the washing step. With any washing procedure, it should be noted that centrifugation may be necessary or of assistance in ensuring complete separation of the layers.

A more elegant and complete though slightly more time-consuming method of removing non-lipid contaminants is to carry out the washing procedure by liquid/liquid partition chromatography on a column. The aqueous washing phase is immobilised on a column of Sephadex G-25 while the organic lower phase is passed through the column. This type of lipid purification procedure was first developed by Wells and Dittmer[573] and has proved useful in a number of laboratories, but a modification of their method described by Wuthier[603] is simpler and more suitable for large numbers of samples. Chloroform, methanol and water in the ratio 8:4:3 (v/v/v) are partitioned as in a conventional "Folch" wash. The column of Sephadex G-25 is packed in the upper of the phases and the crude lipid extract is applied to the column in a small volume of the lower phase. Lipids free of contaminants are eluted rapidly from the column by further lower phase. Gangliosides and non-lipids can be recovered from the column by washing with upper phase and the column regenerated for further use. The procedure can also be used to remove acid and alkali from lipid samples. Siakotos and Rouser[483] have described a more complicated Sephadex column procedure based on a similar principle in which larger amounts of lipids can be purified and in which the gangliosides are also obtained in a discrete fraction free of non-lipids. The procedure appears to be particularly suited to the analysis of bile lipids as the various bile acid components are also obtained in pure fractions separate from the conventional lipids.[446] It is very time-consuming, however, and conditioning and regeneration of the columns is a very lengthy process so that it is not suitable where large numbers of samples have to be purified routinely. When such column procedures are to be used to purify the lipids, it is no longer necessary to stick to the ratio of chloroform–methanol of 2:1 (v/v) in the extracting solvent, particularly as in many instances, chloroform–methanol 1:1 (v/v) may be a better extractant.

4. *Artefacts of extraction procedures*

If chloroform–methanol or indeed any alcoholic extracts which contain lipids are refluxed or stored for long periods in the presence of very small amounts of tissue sodium carbonate or bicarbonate, transesterification of many of the lipids may occur and large amounts of methyl esters are found in the extracts.[313] Similar findings are regularly reported by other workers and it is possible that both acidic and basic non-lipid contaminants may catalyse the reaction. However, as small amounts of methyl esters may occur naturally in tissues, confirmation should be obtained, when they are detected, as to whether they are natural or artefacts of extraction or storage. This can be done simply by extracting the tissues with solvents which do not contain any alcohol such as diethyl ether,[116] hexane[269] or acetone–chloroform[302] and repeating the analysis for methyl esters on this material. Some rearrangement of plasmalogens may also occur when they are stored for long periods in methanol.[561]

Acetone should not be used to extract brain lipids as it causes rapid dephosphorylation of polyphosphoinositides.[112, 575] Acetone extraction of freeze-dried tissue may also result in the *in vitro* production of an acetone derivative (imine) of phosphatidyl ethanolamine.[24, 198]

Although 6-O-acyl-galactosyl diglycerides are known to be natural components of plant tissues,[197] it is possible that they are formed to some extent as artefacts by acyl transfer from other lipids when the cells are disrupted as they are found in much smaller amounts when the tissues are homogenised in the presence of the extracting solvent.[196] Similar difficulties may also be encountered with bacterial homogenates.[480]

5. *Some practical considerations*

All solvents contain small amounts of potential lipid contaminants and should be distilled before use. Plastic containers or apparatus (other than that made of teflon) should be avoided at all costs as plasticisers (usually diesters of phthalic acid) are leached out surprisingly easily and may appear as spurious peaks on chromatograms and affect UV spectra. Wet animal tissues alone in contact with plastic have been reported as extracting small amounts of these compounds and organic solvents will extract very large amounts. Other potential contaminants are discussed in Chapter 3.

Polyunsaturated fatty acids will autoxidise very rapidly if left unprotected in air. Although natural tissue antioxidants such as tocopherols may afford some protection, it is advisable to add an additional antioxidant such as BHT (butylated hydroxy toluene or 2,6-di-*tert*-butyl-*p*-cresol) at a level of 50–100 mg per litre to the solvents.[600] This need not interfere with later chromatographic analyses. Wherever possible, extraction procedures should be carried out in an atmosphere of nitrogen and both tissues and tissue extracts should be stored at $-20°C$ under nitrogen. It is helpful also to deaerate solvents by flushing them with nitrogen before use.

Tissues should be homogenised with solvents in a Waring blender or a similar instrument in which the drive to the knives or grinders is from above so that there is no contact between solvent and any washers or greased bearings. With difficult tissues, clean sand may be added to aid the grinding process. Lyophilised tissues are particularly difficult to extract and it may be necessary to rehydrate them before extraction to ensure quantitative recovery of lipids.

Solvents should be removed from lipid extracts under vacuum in a rotary film evaporator at or near room temperature. When a large amount of solvent must be evaporated, it should be concentrated to a small volume and then transferred to as small a flask as is convenient so that the lipids do not dry out as a thin film over a large area of glass. There is no need to bleed in nitrogen continuously during the evaporation process as the solvent vapours effectively displace any air but the vacuum should eventually be broken with nitrogen. Lipids should not be left for any time in the dry state but should be taken up and stored in an inert non-alcoholic solvent such as chloroform. Last traces of water may be removed by codistillation with chloroform or ethanol.

It may sometimes be advantageous to extract a small sample of the tissue separately to obtain the weight of lipid per g of wet tissue, and also in order that the amount of dry matter in the tissue be determined. The weight of lipid recovered from a given amount of tissue should always be recorded.

B. Recommended Procedures

In the following procedures, it is assumed that all the precautions mentioned above (sections A4 and A5) will be followed.

1. *Extraction of large amounts of tissue*

Where large amounts of tissue have to be extracted and a completely quantitative recovery of lipids is not essential, the procedure of Bligh and Dyer[52] offers some advantages as it does not use as large volumes of solvent as other methods. The following method differs from that originally proposed as it has now been shown that it is advisable to filter the monophasic system before adding water to prevent loss of acidic phospholipids such as phosphatidyl inositol.[393] The yield of lipids is generally 95 per cent or better.

"It is assumed that 100 g of the wet tissue to be extracted contains 80 g of water. 100 g of the tissue is homogenised for 4 min in a Waring blender with a solvent mixture consisting of 100 ml of chloroform and 200 ml of methanol. If the mixture has two liquid phases, more chloroform–methanol should be added until a single phase is achieved. The mixture is filtered through a sintered glass funnel and the tissue residue is rehomogenised with 100 ml of chloroform and filtered once more. The two filtrates are combined, transferred to a 1 l graduated measuring cylinder, 100 ml of 0·88 per cent potassium chloride in distilled water is added and the mixture shaken thoroughly before being allowed to settle. The mixture should now be biphasic (further aqueous solution may be added to ensure this). The upper layer with any interfacial material is removed by aspiration. The lower phase contains the purified lipid and is filtered before the solvent is removed on a rotary evaporator and the lipid is stored in a small volume of chloroform at $-20°C$ for further analysis."

Care should be taken to ensure that the ratio of chloroform–methanol–water is close to 5:10:4 in the monophasic system and 10:10:9 in the biphasic system. A related procedure has been described for tissues that contain a large amount of water, e.g. that of invertebrates.[548]

Tissues which are very rich in lipid such as adipose tissue or oil seeds may be extracted first with diethyl ether or chloroform which will not remove significant amounts of non-lipid contaminants. Last traces of lipid remaining in the tissue can then be recovered by the above procedure.

2. Chloroform–methanol (2:1, v/v) extraction and "Folch" wash

The procedure of Folch, Lees and Stanley[138] is by far the most frequently quoted method for preparing lipid samples from tissues. It has not proved suitable in all circumstances but the modification proposed by Ways and Hanahan[570] is generally satisfactory. This extraction yields approximately a 95–99 per cent recovery of lipids but gangliosides and some of the glycolipids may be lost in the washing step unless the aqueous phase is specifically retained so that they may be recovered.[248] The following procedure is suitable for animal tissues.

"1 g of tissue is homogenised for 1 min with 10 ml of methanol then 20 ml of chloroform is added and the process continued for a further 2 min. The mixture is filtered and the solid residue resuspended in chloroform–methanol (2:1 v/v, 30 ml) and homogenised for 3 min. After filtering, the solid is washed once more with chloroform (20 ml) and once with methanol (10 ml). The combined filtrates are transferred to a measuring cylinder and one quarter of the total volume of the filtrate of 0·88 per cent potassium chloride in water is added, the mixture is shaken thoroughly and allowed to settle. The upper layer is removed by aspiration and one quarter of the volume of the lower layer of water–methanol (1:1) is added and the washing procedure repeated. The bottom layer contains the purified lipid which can be recovered as above."

More extensive, and therefore more time-consuming, extraction procedures will guarantee very little loss, but individual requirements must determine whether the additional effort is worth while. For example, Rouser *et al.*[446] recommend a five-stage extraction procedure in which acidic and basic solvent systems are used. As mentioned earlier, solvent mixtures containing hydrochloric acid[575] or salts[349] are necessary to extract polyphosphoinositides from brain.

Plant tissues must be extracted first with a solvent that inhibits the action of lipases and *iso*-propanol is used most frequently for the purpose. The following procedure has been developed by Nichols.[373, 374]

"The plant tissues are macerated with 100 parts by weight of *iso*-propanol. The mixture is filtered, the solid is extracted again in a similar manner and finally is shaken overnight with 199 parts of chloroform–*iso*-propanol (1:1, v/v). The combined filtrates are taken almost to dryness then are taken up in chloroform–methanol (2:1 v/v) and given a "Folch" wash as above. The purified lipids are recovered from the lower layer as before."

3. *Sephadex G-25 columns for removing non-lipid contaminants*

The procedure originally described by Wells and Dittmer[573] has been widely used and gives excellent results. Wuthier's modified procedure below[603] is somewhat simpler, gives equally satisfactory results and is more suitable for the purification of large numbers of samples. Both methods give good yields of the simpler glycolipids although gangliosides remain on the column to be eluted with the non-lipid contaminants. The crude lipid extracts to be purified are obtained simply by removing the unwashed solvent from the monophasic filtrate after extraction by any of the above procedures.

"Chloroform, methanol and water in the ratio 8:4:3 (v/v/v) are mixed and partitioned. The upper phase (hereafter referred to as UP) and lower phase (LP) are separated and retained. 25 g of Sephadex G-25 (Pharmacia Fine Chemicals, Uppsala, Sweden) are soaked overnight in 100 ml of UP, then are washed with 4 × 100 ml of UP. Columns (1 cm i.d. × 10 cm high) are packed with a slurry of this material. The column is capped with a filter paper disc and rinsed with 10 ml of UP followed by 10 ml of LP. The crude unwashed lipid extract (200 mg), obtained as described above, is taken up on 2–5 ml of LP, filtered to remove any precipitated non-lipid and is applied to the column which is eluted with 25–30 ml of LP at a flow rate of up to 1 ml per minute. The pure lipid is eluted in this solvent and is recovered as before (any UP which leaks through with the eluant does not contain any impurity).

"Gangliosides and non-lipid contaminants can be eluted with 50 ml of UP and the column is regenerated for further use by washing with a further 20 ml of LP."

The complex procedure devised by Siakotos and Rouser [483] can only be recommended in special circumstances, for example in the analysis of bile lipids[446] or for the isolation of gangliosides.[483]

CHAPTER 3

Chromatographic and Spectroscopic Analysis of Lipids. General Principles

A. A Statement of the Problem

Lipid samples obtained from tissues by the methods just described are complex mixtures of individual lipid classes and some means must be devised to obtain each of them in a pure state. Often no single procedure will achieve the desired separations, and combinations of techniques must be used until the required pure lipid classes are obtained. Adsorption chromatographic procedures are generally used to separate each of the various simple lipid classes from the complex lipids. These last may be further fractionated by adsorption chromatography or by ion-exchange chromatography or by combinations of both until the desired separations are attained. During this process, lipid classes may be identified by their reactions with specific chemical reagents, by various spectroscopic techniques or by their chromatographic behaviour relative to that of authentic standards and the amounts of each determined by appropriate methods. The separation of single lipid classes into simpler molecular species is neither necessary nor desirable at this stage. The fatty acid composition of each lipid class can then be determined by gas–liquid chromatography of the methyl esters prepared by transesterification of each and the other hydrolysis products can also be estimated if necessary.

Ultimately, simpler molecular species of lipids may be isolated by partition chromatography, by gas chromatography or by chromatography on adsorbents specifically impregnated with reagents which form complexes with certain functional groups, such as double bonds, in the fatty acid moieties.

In common with all fields of research, the instruments available for lipid analysis are becoming increasingly complex and, therefore, increasingly expensive. None the less, much excellent work can be done with some comparatively simple apparatus. A gas chromatography does not fall into the latter category, unfortunately, and there is likely to be little dispute that very little good lipid research can be carried out without access to such an instrument. Indeed, the vast explosion of information on the chemistry and biochemistry of lipids over the last 10–15 years has been due principally to the development and exploitation of this technique. Apparatus for column or thin-layer chromatography, however, is inexpensive and versatile although sophisticated aids such as automatic fraction collectors and liquid chromatography detectors are extremely useful when available. Infrared and ultraviolet spectrophotometers are likely to be available to most research workers and mass and nuclear magnetic resonance spectrometers, fluorometers and photodensitometers are becoming more commonplace in laboratories. All these refined spectroscopic and other techniques are valued weapons in the lipid analyst's armoury but are not always indispensable to good work; the gas chromatograph is essential.

Greatest attention is given in this book to procedures that utilise gas chromatography and basic column and thin-layer chromatography equipment although techniques which require more complicated instrumentation are not ignored. In the following sections of this chapter, the basic principles of various chromatographic procedures are discussed; specific applications of these are dealt with later in the appropriate chapters.

B. Chromatographic Procedures

1. *Gas–liquid chromatography*

(i) *Principles*

Gas–liquid chromatography (often abbreviated to GLC or GC) is a form of partition chromatography in which the compounds to be separated are volatilised and passed in a stream of inert gas (the mobile phase) through a column in which a high boiling liquid (the stationary phase)

is held on a solid supporting material. The substances are separated according to their partition coefficients which are dependent on their volatilities and on their solubilities in the liquid phase. They emerge from the column as peaks of concentration, ideally exhibiting a Poisson distribution, and are detected by some means which converts the concentration of the component in the gas phase into an electrical signal which is passed to a continuous recorder so that a tracing is obtained with an individual peak for each component as it is eluted. With a suitable detector, the areas under the peaks bear a direct linear relationship to the mass of the components present.

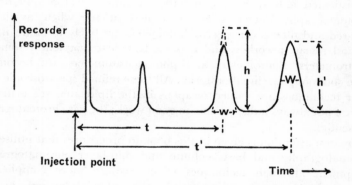

Fɪɢ. 3.1. Schematic diagram of a gas chromatographic recorder trace.

The theory of gas–liquid chromatography has been the subject of a number of excellent textbooks and need not be discussed in detail here, but certain relationships or definitions are particularly useful to the lipid analyst and are worth repeating. Figure 3.1 illustrates a schematic gas chromatography tracing. Shortly after introducing the sample (injection point), there is often a small air peak followed by a peak for the solvent in which the sample was dissolved. The base line should soon stabilise and peaks for the various components begin to energe. The time from the point of the injection to that when the maximum amount of each component is emerging (i.e. to when the peak has reached its maximum height) is known as the *retention time* of the substance. The efficiency of a column is usually expressed in terms of the concept of numbers of *theoretical plates*, originally devised for distillation columns, which can be calculated using the simple formula:

$$n = 16 \times \left(\frac{t}{w}\right)^2$$

where n is the number of theoretical plates, w the width of the base (distance between two tangents drawn from the side of the peak to the base line) and t the retention time (or distance) of the component. The various column parameters such as gas flow rate, length of column and amount of liquid phase are chosen to optimise this figure in order to obtain the maximum resolution of peaks. Although in theory the longer the column, the better the resolution that should be achieved, this is not always so and the desired separations are frequently better obtained by varying the nature of the stationary phase.

One further theoretical relationship is worth mentioning, i.e. for a homologous series of compounds.

Chain length = a constant × log(retention time)

As most lipids occur in nature in such series, this simple formula can be a useful means of provisionally identifying lipid components (see Chapter 5).

(ii) *Apparatus*

A large number of manufacturers now produce gas chromatographs commercially. Individual instruments vary greatly in design and in versatility and the choice of a particular make is dependent on the needs of the individual. All have certain features in common, however. If they are considered in order of contact with the sample, the first of these essential requirements is some means of applying the sample to the column. Usually this is done by injecting the sample, dissolved in an appropriate volatile solvent, by syringe through a silicone rubber septum directly into the packing material of the column ("on-column injection") so that the flow of gas is not interrupted. If the sample is not very volatile, a flash-heater may be used

to ensure that it enters the column entirely as a vapour. The immediate
contact with the liquid phase, which is achieved with on-column
injection, affords some protection to the more labile components and
this method is now preferred in most circumstances. Some commercial
instruments are equipped with an accessory which will automatically
inject samples—a useful feature where large numbers of similar
samples must be analysed.

The columns on which the components are separated are the key
to good analysis and are responsible for much of the versatility of gas
chromatography and are discussed in greater detail below. In many
instruments, there are two identical columns and two detectors which
are balanced against each other to minimise background variations.
The columns are held in an oven, the temperature of which is maintained accurately at the required point within $0.05–0.1°C$. It is frequently
less important that the absolute temperature of the oven be known
precisely than that it be maintained constantly at the set temperature.

Temperature programming is a useful facility on many instruments.
If the sample to be analysed contains components differing widely in
volatility, it is advantageous to start the analysis with the column at a
low temperature so that the more volatile components are separated as
single coherent peaks and then to raise the temperature at a fixed
reproducible rate so that the less volatile components are eluted in a
reasonable time. The sample emerges from the column into the detector
which is maintained at a slightly higher temperature than that reached
by the column by a separate temperature control so ensuring that there
is no change in the response of the detector during temperature
programming.

The flame ionisation detector is that chosen most frequently for lipid
analysis. Eluted components are burned in a flame of hydrogen and air
forming ions that are detected and measured by an electrical system.
Although the mechanism of the ionisation process is not fully understood, this detector is very sensitive, has a good signal-to-noise ratio,
is rugged and does not deteriorate significantly with prolonged use.
It requires a source of hydrogen and air and provided these are constant
during the analysis, the response of the detector to homologous series
of compounds is linearly dependent on the mass of each component
over a very wide range. The flow rate of hydrogen and air together

with that of the carrier gas must be adjusted to the optimum (instructions are generally supplied by manufacturers) for maximum sensitivity. The detector does not respond to some simple carbon compounds such as formic acid or carbon disulphide and the latter is often used as a solvent for samples to be analysed.

Argon ionisation detectors, although potentially more sensitive than flame ionisation detectors, are difficult to maintain in continuous operation as they are very easily contaminated and, as they do not have as wide a linear response range as the flame ionisation detector, they have largely been superseded by this. Thermal conductivity detectors, though simple in theory and in use, are much less sensitive than most other detectors but have some advantages for preparative gas chromatography as the sample passes through the detector unchanged.

There are of course many other features of importance in choosing a gas chromatograph, including sensitive gas flow controllers and the quality of the amplifier and recorder supplied with the instrument and all such factors must be considered. Only the highest purity gases should be used. Helium is probably the best carrier gas but is prohibitively expensive outside the U.S.A. and nitrogen containing less than 5 ppm of oxygen is most often favoured as an alternative. Argon is more expensive than nitrogen but usually contains less oxygen and has advantages for the chromatography of more labile compounds. Containers of molecular sieves inserted into the gas lines to remove impurities improve the signal-to-noise ratio on the gas chromatographic traces.

(iii) *Columns*

The heart of the gas chromatograph is the column on which the substances are separated. There are three basic types: packed, open-tubular (capillary) and support-coated open-tubular (often abbreviated to SCOT).

Packed columns are the work horses of gas chromatography. They consist of glass or metal tubes, generally $\frac{1}{4}$ inch or $\frac{1}{8}$ inch in diameter (o.d.) and 4 ft to 7 ft in length, coiled to fit the oven unit and filled with an inert solid support coated with the liquid phase. Stainless steel is the most popular material for the columns themselves because of its durability and because it is remarkably inert chemically to lipids. Aluminium is occasionally chosen for the purpose also, but copper columns are not suitable because of strong adsorptive effects towards lipids. All metal columns must be thoroughly cleaned with solvent before use to remove any grease, and it may be advantageous to remove active sites by silanisation. Glass columns are more fragile but otherwise have a number of advantages over metal columns; the surface is almost completely inert, it is easy to see how efficiently the column has been packed and whether any breaks in the packing appear in use while deterioration of the top of the column packing at the inlet end is immediately apparent. Also, they are more easily emptied for reuse. Now that columns with good glass to metal seals are commercially available, glass columns should become more popular especially where the columns remain *in situ* in the gas chromatograph for long periods as breakages are most likely to occur on removing or replacing columns in the oven unit.

The list of liquid phases available to the analyst is almost endless but, in practice, certain polyester liquid phases have proved themselves most useful for fatty acid analysis and a few silicone elastomers have advantages for higher molecular weight components. These will be discussed in greater detail in later chapters where analyses of specific lipid classes are described. The solid supporting materials for the liquid phases are generally diatomaceous earths, graded so that the particles are of uniform size and deactivated by acid washing to prolong the life of the liquid phases and by silylation to minimise any adsorptive effects on the solutes.

It is claimed[227] that the most uniform coating of the liquid phase on the support are obtained if the liquid in a solvent is filtered through a bed of the support and the whole spread out as a layer to dry; the amount of liquid phase that remains on the support must be determined by experiment and depends somewhat on the nature of the support. It is also possible to achieve satisfactory coatings by evaporating a solution of the phase in the presence of the support in an indented flask (to stir the solid material) on a rotary evaporator but great care must be taken with this method to ensure that the support is not damaged. Precoated supports are now commercially available at prices close to that of the starting materials. Columns are packed by adding the coated support in small amounts to the column while tapping gently and applying a vacuum to the exit end of the column. When the column is filled, a glass-wool plug (acid-washed and silanised) is placed on top of the packing at the inlet end to consolidate it. If the columns are too loosely packed, the separations are poor; if too tightly packed, they may block or the injection syringes may plug easily. The column must be conditioned for up to 48 hr at a temperature just above that at which it is to be operated before being used. 5–7 ft columns containing a support with 10–15 per cent (w/w) liquid phase should have an efficiency of 3000–5000 theoretical plates. Such columns normally have a working life of well over a year and this can frequently be prolonged by repacking the top inch or two with fresh packing material periodically.

Open-tubular or capillary columns (comprehensively reviewed by Ettre[132]), as the name suggests, consist of lengths of up to 100 m of narrow bore stainless steel (or occasionally glass) tubing, normally 0·01 inch in internal diameter, the inner wall of which is coated with the liquid phase. Such columns are highly efficient, 20,000–100,000 theoretical plates overall, and give quite remarkable separations; for example, resolution of optical enantiomers or of positional isomers of monoenoic fatty acids may be possible. They are particularly useful when coupled directly to the inlet of a mass spectrometer as each peak is likely to contain only a single component so that more positive identifications can be made. Capillary columns have a number of serious disadvantages, unfortunately; lengthy analysis times are necessary and the prolonged exposure to the metal surface results in decomposition of polyunsaturated components, coating the columns with liquid phase is difficult and not always reproducible, the working life of the columns is very short, often only a few weeks and they are difficult to repack. Also, they will only accept a very small sample for analysis and an elaborate stream splitting device is necessary at the injection port to ensure that the column is not overloaded and this accessory is not available with all commercial gas chromatographs. Capillary columns then tend to be used only for a few specialised applications.

Support-coated open-tubular columns[133, 413] consist of wider bore tubing than the capillary columns (0·02–0·05 inch in internal diameter) and contain a finely powdered solid support coated with the liquid phase. They are usually shorter in length than the capillary columns

(10–15 m) and the efficiencies are a little less (10,000–15,000 theoretical plates). They can, however, take much larger loads of sample than the capillary columns, the analysis time is shorter, the operating temperature is lower and the useful working life is longer. Such columns cannot easily be repacked, however, and are too expensive for routine use.

(iv) *Quantification of components*

It appears to be generally accepted that the most accurate method of estimating the amount of material in a gas chromatographic peak is by electronic digital integration of the amplified signal from the detector and this is especially true for temperature-programmed analyses. A number of sophisticated electronic integrators are available commercially specifically for the purpose but the detector output can also be fed into a digital voltmeter and then via suitable software into a computer that has been programmed for the analysis.[58, 81] Such equipment is of course extremely expensive and not normally available to the analyst.

The alternative is to measure the area of the peaks or some parameter proportional to this by hand. Planimetry or measuring the area of the triangle enclosed by tangents drawn to the sides of the peaks to the base line have been used but are not particularly accurate methods, i.e.

$$\text{Area} \propto h \times w \quad \text{(Fig. 3.1)}$$

A better procedure is to multiply the height of the peak by the width at half-height.

$$\text{Area} \propto h' \times W \quad \text{(Fig. 3.1)}$$

If the peaks obey a pure symmetrical Poisson distribution, the height of the peak multiplied by its retention time is also proportional to the area.[41]

$$\text{Area} \propto h \times t' \quad \text{(Fig. 3.1)}$$

Although peaks are never completely symmetrical, there have been such marked improvements in the design of gas chromatographs and in the quality of the inert supports and liquid phases over the last few years, that this method of quantification is more accurate for isothermal analyses than reports in the earlier literature might lead one to believe. It has the advantage that two large distances are measured so that the measuring errors are smaller, it gives reasonably accurate results with incompletely resolved peaks and it is the simplest and most rapid of the methods. A correction factor can be applied to increase the accuracy of the method by compensating for the fact that the sample is applied to the column as a finite band rather than as a point source,[59] i.e. the widths of the peaks at half-height are plotted against the retention times of components and the straight line through these intercepts the base line at a point in front of the point of the actual injection. If retention distances are measured from this point, more accurate results can be obtained, particularly for early running components. This procedure cannot be used with temperature programmed analyses, however. Usually the areas of the peaks, determined by one of the above methods, are summed and the area of each is expressed as a percentage of the total so that the proportion of each component in the mixture is obtained.

All methods of peak area measurement should be calibrated with standard mixtures of known composition, similar to the samples to be analysed, and the calibration should be checked at regular intervals. It may then be necessary to adjust the area measurements by incorporating response factors obtained from the calibrations into the calculations. Detectors usually respond to the weight of a given component present and weight percentages may have to be converted to molar percentages by multiplying the results by suitable arithmetic factors. If the above precautions are taken, an absolute error of less than 1 per cent of major components (i.e. 10 per cent or more of the total) of any mixture should easily be achieved.

(v) *Range of applications*

Gas–liquid chromatography is used to analyse most substances that can be volatilised without decomposition. The highest molecular weight compounds that have been subjected to the technique are triglycerides containing over sixty carbon atoms, which can be eluted from highly thermostable silicone liquid phases at temperatures approaching 350°C. Although this is close to the cracking temperature of such compounds, the liquid phase appears to offer some protection.

C

It is unlikely that compounds of very much higher molecular weight will ever be successfully subjected to conventional gas chromatography but instruments are under development which operate at very high pressures (100 atm or more) and these may produce exciting new applications.

The most useful application of gas chromatography in lipid analysis is in the determination of the fatty acid compositions of lipids and by a judicious choice of liquid phases, a complete analysis of fatty acids separated both by chain length and degree of unsaturation in a given sample can be achieved. Indeed, positional isomers of unsaturated fatty acids can also be separated on open-tubular columns. Volatile derivatives of lipids must often be prepared for GLC analyses: for example, fatty acids are usually converted to methyl esters and the free hydroxyl groups of partial glycerides are acetylated. In addition to fatty acid analysis, gas chromatography can be used to determine cholesterol, glycerol, inositol, carbohydrates and many other compounds, released on hydrolysis of lipids, in the form of volatile derivatives.

Although it is primarily an analytical tool, gas chromatography may also be used preparatively by inserting a stream splitter between the end of the column and the detector and collecting the effluent in a suitable trap. Gas chromatographs can also be operated with the detectors in series or in parallel with suitable counting equipment so that mass and radioactivity measurements on compounds eluted from the columns are determined together in a single analysis. A mass spectrometer coupled to the exit of a gas chromatograph is a very powerful tool for determining the structures of unknown compounds. These and other applications will be discussed in detail in the appropriate chapters.

2. *Adsorption chromatography*

Adsorption chromatography utilises differences in the degree to which lipid components are adsorbed on to a solid support as a means of separating them. Lipids are held by the adsorbents in a variety of ways including by hydrogen bonding, by van der Waals' forces and by ionic bonding. The more polar the functional groups contained in the lipid, the more strongly is it adsorbed; fatty acid chains are non-polar and have little effect in relation to polar functional groups such as free hydroxyl or keto groups or the polar head-groups of complex lipids. Lipids are released from the adsorbent by passing solvents of increasing polarity through it and are therefore separated by such procedures according to the number (but generally not kind unless they contain polar functional groups) of fatty acids in the molecule and the number and kind of other polar functional groups. The adsorbent may be held in glass columns and eluted continuously with solvents (column chromatography) or it may be supported as thin layers on glass plates and the solvent allowed to pass through the adsorbent by capillary attraction (thin-layer chromatography, often abbreviated to TLC). The principles of adsorption chromatography have recently been reviewed in some detail by Stein and Slawson.[517]

(i) *Silicic acid column chromatography*

By far the most widely used adsorbent is silicic acid, a partially hydrated silicon dioxide (also often termed silica or silica gel) and it is sold commercially by a number of manufacturers. Different brands and even different batches of the same brand may vary in such properties as particle size, degree of hydration, surface area and in the proportions of trace organic and inorganic contaminants. In general, the finer the particles of silicic acid, the greater the surface area available for adsorption and the better the separations that can be obtained. If the particles are too small, however, the flow rate through the column may be too slow so, in practice, a compromise must be sought which allows reasonable flow rates without unduly prejudicing the quality of the separations. 200-mesh silicic acid has been commonly used but some manufacturers now market products prepared by special means to have larger particle sizes without sacrificing surface area and these permit more rapid flow rates to be attained.

The degree of hydration of the silicic acid also has an important influence on the quality of the separations. If there is too little water, lipids are strongly held and are not eluted from the column in sharp bands but tend to "tail". If the adsorbent contains too much water, lipids are not adsorbed sufficiently strongly and the separations are poorer. The nature of the separation may also change in this instance as partition effects between bound water and polar lipids may be observed. Although the optimum degree of hydration must be found by trial and error, it is generally around 5 per cent. Water can be removed from the adsorbent by heating it at 110°C for several hours or by prewashing the column with dehydrating solvents such as acetone.[204] Water can be added uniformly to an adsorbent by adding the desired amount to the dry powder in a stoppered flask, turning this slowly until any large lumps break up and leaving for about 2 hr to allow the water to distribute evenly.

Glass columns with a sintered disc at the bottom to hold the adsorbent and a tap below this so that the flow of solvent can be controlled are suitable for column chromatography (Teflon taps are preferable when lengthy separations are involved) but the dead volume should be as small as possible. Highly sophisticated columns with very small dead spaces at the top and bottom are commercially available but these are more elaborate and expensive than is necessary for most routine lipid separations. The adsorbent is packed in a slurry of the first solvent to be used in the separation and is allowed to settle with gentle tapping until it forms a bed at least ten times greater in height than in diameter (long thin columns give better separations than short wide ones). Columns should never be allowed to run dry as channels are formed that cause uneven elution of lipids.

The amount of lipid that can be applied to a column is variable and depends on the magnitude of the differences in polarity between the various components to be separated. In general, 30 mg of lipid per g of adsorbent is a reasonable load but this can be varied with circumstances. The sample should be applied to the column in as small a volume as possible of the least polar eluting solvent and washed on to the bed of adsorbent carefully until no lipid remains above the surface when the main solvent reservoir can be attached.

Components are removed from the column by eluting with solvents of gradually increasing polarity. Wren[599] has described a modification of Trappe's eluotropic series of solvents. This is, in order of increasing polarity, petroleum ether (hexane) < cyclohexane < carbon tetrachloride < benzene < chloroform (containing no ethanol as stabiliser) <

dichloromethane < diethyl ether < chloroform stabilised with 1 per cent ethanol < ethyl acetate < acetone < acetonitrile < methanol < acetic acid < water. Solvents can be changed in a discontinuous manner (stepwise elution), which is simple in practice and suited to many of the more common lipid separations, or they can be changed continuously (gradient elution) by means of several simple devices.[204, 578] Although the merits of the two types of elution procedures have been widely debated, no clear favourite has emerged. It appears that the commoner phospholipid classes may be better separated by stepwise elution but that gradient elution may give better results with cerebrosides and sulphatides. Both methods may have to be tried with difficult samples.

The quality of the separations is also dependent on the flow rate of the eluting solvent and again the optimum may have to be determined by trial and error though flows of 1–3 ml/min usually give satisfactory results. With very tightly packed columns, it may be necessary to apply pressure to the top from a suitable pump or from a nitrogen cylinder to obtain satisfactory flow rates. This is preferable to applying suction from below which causes the column to dry up at the bottom.

Column chromatography with silicic acid offers some advantages over TLC as the lipids are better protected against autoxidation (although there is a recent report[495] that lipids are better protected on thin-layer adsorbents than had hitherto been believed) and as larger quantities can be separated. Partial glycerides will isomerise on silicic acid but most other lipids are unaltered in its presence.

(ii) *Other adsorbents*

Florisil, which is the trade name for a coprecipitated mixture of magnesia and silicic acid produced by the Floridin Co. (Pittsburg, U.S.A.), offers some advantages over silicic acid in the separation of simple lipids.[73] It is supplied as a coarse mesh powder with a high adsorptive capacity for lipids so that large amounts of these may be separated at higher solvent flow rates than are normally permissible. While the separations that can be achieved are similar in many ways to those obtained with silicic acid, acidic lipids are very strongly adsorbed and must be eluted with acidic solvent systems. Recoveries of phospholipids are generally poor, especially of phosphatidyl choline, and as some magnesium silicate is eluted with polar solvents, it may be necessary to purify complex lipids obtained in this way by a "Folch" wash (see Chapter 2).

When Florisil is washed with concentrated hydrochloric acid, much of the magnesium is removed and the product is in effect a coarse mesh silicic acid with a similar adsorptive capacity to the original Florisil but with identical chromatographic properties to silicic acid.[74] The high flow rates obtained with this product render it suitable for rapid routine separations. It is prepared as follows:

> "Florisil (300 g; 60–100 mesh) is mixed with concentrated hydro-chloric acid (900 ml) and heated on a steam bath for several hours. The hot supernatant liquid is decanted and the adsorbent is washed with a little more acid and heated overnight with a further 900 ml of acid. The product is filtered and washed till neutral with distilled water and the acid and washing treatment repeated. The neutral residue is washed with approximately 400 ml each of methanol, chloroform–methanol (1:1, v/v), chloroform and diethyl ether. The product is first air-dried and then activated by heating over-night at 110–120°C."

Alumina has also been used as an adsorbent for column chromato-graphy but there have been reports of extensive alteration to lipids including autoxidation of double bonds, isomerisation of partial glycerides and hydrolysis of lipids when it is used. Different batches of alumina appear to vary more in properties than does silicic acid so for most purposes the latter is now preferred.

(iii) *Thin-layer chromatography*

In thin-layer chromatography (TLC), the adsorbent is held on glass plates in a thin layer as the name suggests. A very fine grade silica gel is by far the most common adsorbent used for the purpose and this may contain calcium sulphate as a binder so that the layer adheres to the plate. Such mixtures are commonly termed "silica gel G" although, strictly speaking, this is the trade name of a commercial product. The adsorbent is applied to the plate in the form of an aqueous slurry (2 ml of water per g of adsorbent) by means of a suitably designed spreader so that even layers of a predetermined thickness are obtained. The plates are air-dried briefly then activated by heating in an oven at 110–120°C for 2 hr and stored in an airtight box or in a desiccator. A number of manufacturers supply complete kits of equipment at moderate cost for preparing TLC plates, but, if it is intended to work with silver nitrate impregnated layers (see later section), the spreader should be made of anodised or silver-plated aluminium or some other inert material so that it is impervious to the reagent. While it is also possible to purchase pre-coated TLC plates ready for immediate use, they are generally too expensive to be used routinely. Small TLC plates suitable for simple separations can be made by dipping microscope slides in a slurry of silica gel in chloroform (0.25 g per ml) and allowing them to dry in the air[395] or microspreading equipment is available. Alternatively, it can be economic to cut up commercial pre-coated layers (on plastic backing) into small pieces.

Samples are applied to the plate in a solvent (frequently chloroform) by means of a syringe or with a sample applicator made specifically for the purpose (many are available commercially) as discrete spots or as narrow streaks 1·5–2 cm from the bottom of the plate, and the plate is then placed in a tank containing the eluting solvent. Lining the tanks with filter paper to saturate the atmosphere inside with solvent vapour speeds up the analysis and many occasionally improve the resolution. The solvent moves up the plate by capillary action taking the various components with it at differing rates according to the extent to which they are held by the adsorbent. When the solvent nears the top of the plate, the plate is removed from the tank, dried and sprayed with a reagent that renders the lipids visible. These should appear as a vertical line of discrete spots or bands.

The spray may be a chemical reagent which is specific for certain types of lipid or for certain functional groups (see later chapters) or it may be a non-specific reagent that renders all lipids visible. A 0·1 per cent (w/v) solution of 2′,7′-dichlorofluorescein in 95 per cent methanol is most frequently used for the latter purpose and causes lipids to show up as yellow spots under UV light. Alternatively, an aqueous solution of Rhodamine 6G (0·01 per cent, w/v) may serve the purpose in which case lipids appear as pink spots under UV light. Rhodamine 6G is particularly useful when alkaline solvent systems have been used and 2',7'-dichlorofluorescein is to be preferred with acidic solvents. Both sprays are non-destructive to lipids which can be recovered from the plates for further analysis. Water can be used as a non-destructive spray when large amounts of lipids are separated preparatively; they show up as white spots on a translucent background. Lipids also become visible as brown spots if the plate is left for a few minutes in a tank of iodine vapour, but the iodine reacts to some extent with polyunsaturated fatty acids which cannot then be recovered for analysis in other ways. Alternatively, the plates may be sprayed with a solution of 50 per cent sulphuric acid and the lipids made visible as a black deposit of carbon by heating the plates at 180°C for an hour or so. 20 per cent (w/v) ammonium bisulphate in water has also been used to char lipids[55] and the vapours given off during the heating process are less corrosive than those when concentrated acids are used. Although charring procedures have the obvious disadvantage that they completely destroy the lipids, they are very sensitive and as little as 1 μg of lipid can be detected by this means. Sterols give a red-purple colour in a few minutes with charring reagents before blackening and this is a useful diagnostic guide.

For analytical purposes, layers of adsorbent 0·25 mm thick or less give maximum resolution, but in preparative applications thicker layers are necessary to take heavier loads of lipids. The thicker the layer, the poorer the resolution and, in practice, a compromise must be sought and layers 0·5–1·0 mm thick are usually preferred; indeed with thicker layers, there may be mechanical difficulties in keeping the adsorbent adhering to the plate. The amount of lipid that can be applied to a preparative TLC plate varies with the ease of separation of the components of the mixture. For example, 25–50 mg of simple lipids can often be applied to a 20×20 cm plate with a layer 0·5 mm thick of silica gel G. On the other hand, only 10 mg or so of phospholipids can be separated effectively on such plates.

Lipids separated by TLC can be recovered from the plates after they have been detected by an appropriate non-destructive method by scraping the adsorbent band into a small chromatographic column or sintered disc funnel and eluting with solvents of sufficient polarity to remove the lipids. Chloroform or diethyl ether containing 1–2 per cent methanol (by volume) will elute most simple lipids, although partial glycerides may isomerise in alcoholic media, and chloroform–methanol–water (5:5:1 by volume) will quantitatively elute most polar lipids. Lipids can also be recovered by repeatedly mixing the adsorbent with solvent in a test tube, centrifuging and decanting the supernatant liquid. The dye used to detect the lipid is also eluted by these procedures and can be removed from non-polar lipids by washing them in diethyl ether or chloroform through a short column of Florisil. 2′,7′-Dichlorofluorescein can be removed from non-acidic polar lipids by dissolving them in chloroform–methanol (2:1 by volume) and giving them a "Folch" wash with a solution of tris buffer (0·05 M) of pH 9 or with dilute ammonia. The dye does not necessarily interfere with subsequent analyses: for example, if the fatty acid components are transesterified for GLC analysis, the dye is not eluted from the GLC column.

Complicated lipid mixtures cannot always be separated by thin-layer chromatography in one direction but can often be resolved by rechromatography in a second direction (two-dimensional TLC). In this method, the sample is applied to the plate as a spot in the bottom left-hand corner of a square TLC plate and the plate run normally in a selected solvent system. When the solvent has run close to the top of the plate, the plate is removed from the tank and dried thoroughly in a desiccator under vacuum so that atmospheric moisture is not permitted to deactivate the adsorbent. The plate is then turned anti-clockwise through 90° and redeveloped with a second solvent system which is normally different from the first (some workers apply the lipid to the bottom right-hand corner and turn the plate clockwise).

As in column chromatography, the quality of the separations and the distances that individual lipid classes will migrate is influenced by the degree of hydration of the adsorbent. This is affected by the time and temperature of activation of the plates, by the storage conditions and, as plates are inevitably exposed to the air during the application of samples and during development, by the relative humidity of the atmosphere. Different proprietary brands of silica gel and different batches of the same brand have different particle sizes and surface areas and, therefore, different adsorptive properties and this also affects the nature of the separations. If these factors are very rigorously controlled, reproducible Rf values for different lipid classes can be obtained but this is rarely worth the effort involved. A much more common practice is to apply a mixture of authentic lipid standards, that migrate in a known order, to the plate alongside the unknown mixture so that direct comparison is possible.

TLC offers a number of advantages over column chromatography. It is more rapid and sensitive, gives better resolution and the separation can almost instantly be seen. This last feature is particularly useful in preparative applications as, if the desired separation is not achieved with one development, the plate can be given a second or third development in the same direction with the same or a different solvent often resulting in an improved resolution. Lipids appear to be more stable to autoxidation on thin layer adsorbents than has generally been believed.[495] None the less, it is advantageous to add antioxidants such as BHT to the sample, to the eluting solvent or to the spray reagents (BHT migrates with the solvent front even in non-polar solvent systems) to protect the lipids during subsequent analyses. Although larger amounts of lipid can be purified or separated in a single operation by column chromatography, because of the difficulty of monitoring column eluants, it may often be quicker and less effort to separate lipids preparatively by chromatography on several identical TLC plates, combining corresponding fractions.

Silicic acid impregnated paper[134, 263] is occasionally used as an alternative to TLC and similar though not identical types of separation are obtained with the two procedures. It is claimed that the impregnated papers are superior in that a greater variety of spot tests can be applied to a given chromatographic paper, that developed and stained papers are more easily stored than TLC plates and that autoradiography is simpler. On the other hand, TLC is a more versatile technique and in particular sensitive charring reagents can be used to detect lipids and much greater amounts of material can be separated preparatively.

(iv) *Quantification of components*

Lipids in general do not possess chromophores or functional groups that can be readily detected spectrophotometrically so that until recently it has not been possible to monitor the effluents of liquid chromatographic columns continuously. Sophisticated detection systems are now available commercially, however, that can be used for the purpose. A clean stainless steel wire is passed continuously through the column effluent, then into an oven to dry off excess solvent and finally into a very hot oven to volatilise or pyrolyse the eluted lipids. These are flushed into a flame ionisation detector, the signal from which is monitored continuously by an electronic recorder so that eluting lipids appear as a series of peaks on the trace. Such equipment is naturally very expensive and not always available to the analyst.

In the absence of this equipment, the eluant from columns is collected in small volumes (10–20 ml) with a fraction collector and aliquots of constant volume are taken from each of these (or from every 2nd or 4th) so that the amount of lipid can be determined by some means. For example, the solvent can be removed and the lipid weighed on a microbalance. Usually the solution is placed on a small aluminium pan weighing 5–10 mg which is dried on a hot plate at 60°C for a few minutes then in a vacuum desiccator to constant weight. The pans should never be touched by hand as sufficient grease may be deposited to alter the weight significantly. As the lipid autoxidises during the process, the pans are discarded after weighing. Such gravimetric procedures are best reserved for when large quantities of material are being chromatographed.

In an alternative method,[19, 139, 489] the solvent is removed from an aliquot of each fraction of the eluant and 2 ml of a reagent prepared by dissolving 2·5 g of potassium dichromate in 1 l of 36 N sulphuric acid, is added. The whole is heated at 100°C for 45 min with shaking, cooled and the decrease in the absorbance of potassium dichromate ($Cr_2O_7^=$) at 350 nm is determined. The lipid is oxidised and the dichromate reduced by a proportionate amount that can be determined from the absorbance measurements. Appropriate blanks must be run and the procedure calibrated with authentic standards. The procedure is useful for neutral lipids but less so for phospholipids which are best estimated by a direct phosphorus determination by the method described in Chapter 7. It is also possible to apply aliquots of the eluant as spots to TLC plates—25 or more fractions per 5×20 cm plate—and char these by spraying with chromic acid in sulphuric acid (6 g of potassium dichromate per litre of 55 per cent sulphuric acid) and heating at 180°C for 30 min.[410] The amount of material in each spot

Lipid Analysis

can be estimated roughly by eye or more accurately by photodensitometry. Again, the absorbance of blanks must be determined and the
method calibrated with authentic standards. TLC with microplates
may be useful for identifying and estimating the purity of fractions
from columns.

When the adsorption characteristics and the lipid elution profile of a
silicic acid column have been determined, it may no longer be necessary
to monitor fractions so precisely.

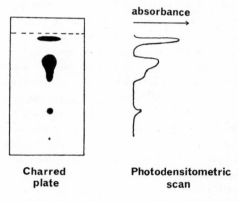

| Charred | Photodensitometric |
| plate | scan |

Fig. 3.2. Charred TLC plate and corresponding photodensitometric scan.

Charring procedures followed by photodensitometry can also be used
to estimate lipids separated by TLC. Bands of lipid on the adsorbent
are scraped into tubes to which the dichromate reagent described
above is added. The tubes are heated as before, centrifuged to precipitate the silica gel and the absorbance of the solution is measured at
350 nm.[19, 139, 489] Alternatively the plate can be sprayed with chromic
acid solution[443] or 3 per cent cupric acetate in 8 per cent phosphoric
acid[137] and the amount of charred material obtained after heating the
plate at 180°C for 25 min measured by means of a scanning photo-
densitometer as illustrated in Figure 3.2. The areas of the peaks on the
recorder trace are proportional to the amount of lipid originally
present. This procedure has a number of disadvantages: the sample is

destroyed, the yield of carbon is variable, authentic standards are necessary for calibration but may not always be available and there are considerable doubts as to how linear the relationship is between the density of a spot and the amount of material in it. In addition, a scanning photodensitometer is an expensive specialised piece of equipment and is not to be found in all laboratories. However, where large numbers of similar samples have to be analysed routinely, this is probably the best method. Phospholipids on silica gel can be estimated by phosphorus determination (see Chapter 7) in the presence of the adsorbent. This and the charring procedure may give high blank values in the presence of silica gel, so reducing the accuracy of such methods for estimating small amounts of lipid or minor components of a given mixture, but the problem can be minimised by thoroughly extracting the adsorbent with solvents before making up the TLC plates. For example, silica gel without binder can be washed in a Buchner funnel with 8 volumes of chloroform–methanol–formic acid (2:1:1 by volume) followed by 4 volumes of distilled water. The adsorbent is finally dried in an oven at 110°C for 48 hr with occasional stirring.[394] Alternatively, plates can be developed to the top with a solvent such as diethyl ether–methanol (1:1, v/v) then used again in the normal way after reactivation (they must of course be used in the same direction).

A more promising development lies in the estimation of lipids by fluorometry[201, 379, 440, 459] in which the fluorescence of a dye, produced by the presence of lipid, is measured. The dye can be incorporated into the adsorbent,[440] or sprayed on to the developed plate[201, 379] and the fluorescence measured by a scanning fluorometer or the lipids can be recovered from the plates, the dye added in an appropriate solvent and the fluorescence measured by placing the mixture in cuvettes in a fluorometer.[459] The fluorescence is directly proportional to the amount of lipid present, but depends on the nature of the lipid, so calibration curves must be prepared with standards. As double bonds in alkyl chains cause quenching, these standards should be very similar in nature to the samples to be analysed. Standards should also be run on the same plate as the unknowns to compensate for differences in quenching obtained with different batches of adsorbent. Rhodamine 6G (0·01 per cent in water) has been used most often as the dye but a spray of 1-anilino-8-naphthalene sulphonate (ANS) at the same concentration has also been used.[201] In a typical analysis, spots of the lipids to be determined and standards are placed in a row on the plate which is developed in the normal way then dried thoroughly and sprayed evenly with the dye. It is then dried thoroughly again and the individual lanes of lipid on the plate are scanned with the fluorometer; the amount of each lipid is obtained from the calibration curve after making an allowance for quenching. The procedure is non-destructive so that the lipids can be recovered for further analysis if necessary and it has been applied to the determination of both simple and complex lipids although the range of lipids tested is still somewhat limited and the technique has not been used in many laboratories. It obviously holds great promise for the future, however.

Gravimetric methods are very unreliable for estimating lipids separated by thin-layer chromatography as small amounts of impurities, including calcium sulphate, silica gel and the indicating dye may also be eluted from the plates and weighed.[280]

As all lipids contain fatty acids, it is possible to determine the amounts of lipid classes separated by chromatographic procedures by determining the amounts of the fatty acids that they contain. Chemical methods are available for the purpose (see Chapter 6), but gas chromatography can also be used. Typically, the fatty acid components of each lipid are converted to methyl esters in the presence of a known amount of the ester of an acid that does not occur naturally in the sample (e.g. an odd-chain compound) and this serves as an internal standard. Transesterification can be performed on pure lipids eluted from TLC plates or it can be carried out in the presence of the adsorbent, but the method chosen should be such that no losses can occur before all the fatty acid components are in the form of methyl esters (see Chapter 4). By means of gas chromatography, the total amount of the acids relative to that of the standard is obtained by dividing the sum of the areas of the fatty acid peaks on the recorder trace by that of the internal standard (see Fig. 3.3.) This method has been widely used to determine the amounts of molecular species of single lipid classes[172] but can also be used to estimate natural mixtures of lipids;[99] it has the additional merit that both the fatty acid compositions and the amounts of the lipid classes in a given mixture are determined in a single analysis. Related procedures are used in the estimation of intact

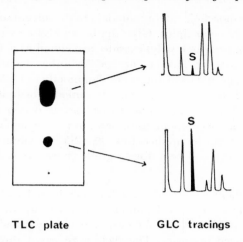

TLC plate   **GLC tracings**

FIG. 3.3. The use of a fatty acid internal standard to estimate, by means of GLC, lipids previously separated by TLC (peak "S" is the internal standard).

glycerides, cholesteryl esters, cholesterol and many of the hydrolysis products of lipids such as glycerol or carbohydrates and these applications are discussed in greater detail in later chapters. Methods of this kind are not always suited to the routine analysis of large numbers of samples, however.

3. *Diethylaminoethyl (DEAE) cellulose chromatography*

DEAE cellulose chromatography is a useful technique for the separation of complex lipids in comparatively large amounts. The principle of the separation process is partly ion-exchange chromatography of the ionic moieties of the polar lipids and partly adsorption of highly polar non-ionic parts of complex lipids, for example the hydroxyl groups of inositol or carbohydrates. The DEAE cellulose packing material is converted to the acetate form from which charged non-acidic lipids (such as choline or ethanolamine containing lipids) are eluted with chloroform containing various proportions of methanol. Weakly acidic lipids are eluted with glacial acetic acid and it is necessary to add an inorganic salt or ammonia to the solvent to elute strongly acidic or highly polar lipids. Glycolipids can also be separated by the procedure and ceramide polyhexosides are eluted as a single complex fraction that can be further purified by other techniques. DEAE cellulose in the borate form and triethylaminoethyl (TEAE) cellulose[446] are also useful for some purposes. The amount of lipid that can be applied to such a column without overloading it depends somewhat on the nature of the components present, but as a rough guide approximately 300 mg can be applied to a $30 \times 2 \cdot 5$ cm column. The DEAE cellulose, which should be a fibrous grade (fines reduced), is prepared as follows:[444]

"The adsorbent is washed on a Buchner funnel with 1 N aqueous hydrochloric acid (3 bed volumes) followed by water to neutral pH and then with $0 \cdot 1$ N aqueous potassium hydroxide (3 bed volumes) followed by water. The cycle is repeated three times. It is converted to the acetate form by washing with 3 bed volumes of glacial acetic acid and by leaving overnight in this solvent. The column is packed with the slurry of DEAE cellulose in glacial acetic acid, adding small amounts of material at a time and tamping down gently with a wide glass rod. If fines generated by the washing procedure are removed by decanting the upper layers of the slurry two or three times before packing the columns, better flow rates are obtained. The bed is freed of acid by washing with 5 bed volumes of methanol, 3 bed volumes of chloroform-methanol (1:1 v/v) and 5 bed volumes of chloroform. Non-polar coloured compounds such as azulene can be run through the column to determine the bed volume and to detect imperfections in the packing."

Lipids eluting from the columns may be determined by the procedures described above for adsorption column chromatography.

4. *Partition chromatography*

In partition chromatography, lipids are separated by means of differences in their partition coefficients between two immiscible liquids. It is especially useful for separating lipids with homologous or vinologous series of aliphatic residues and in particular for the isolation of fatty acids or their esters. A *cis*-double bond has approximately the same effect on the partition coefficient of a lipid as two methylene groups; thus palmitic and oleic acids or their derivatives are difficult to separate and are known as a *critical pair*. Also, *cis*- and *trans*-components are not readily separated, but a triple bond is equivalent to more than two double bonds and polar functional groups will drastically alter the partition coefficient of a compound. Where the compounds to be separated differ markedly in their partition coefficients, for example normal fatty acid esters and hydroxy acid esters or neutral lipids and phospholipids, useful separations can be achieved with a few transfers in separating funnels. For more difficult separations, such as simple fatty acid homologues, several hundred transfers on a specially constructed countercurrent distribution machine are necessary. This is probably the mildest separatory procedure available to the lipid chemist although it is tedious and time-consuming and the equipment is expensive. It has proved particularly useful for the isolation of labile fatty acids from seed oils.

Chromatographic procedures utilising this phenomenon are more convenient in practice. In *partition chromatography*, a polar solvent (the stationary phase) is held on an inert support on a column or on a TLC plate and a less polar solvent (the moving or mobile phase) flows continuously past it. Paper chromatography separates compounds by this principle as in this instance, the stationary phase is water bound to the cellulose. In *reverse phase partition chromatography*, the non-polar solvent is held on the inert support while the polar solvent moves. Useful systems utilising both principles have been described for separating fatty acids and esters and for molecular species of triglycerides and phospholipids but reverse phase systems are generally more useful.

In reverse phase column chromatography, the non-polar stationary phase is held on an inert support such as silanised celite[232] or Kieselguhr while the mobile phase, in which the sample is applied, is percolated through the column. The stationary phase is generally a hydrocarbon such as heptane or a high boiling silicone liquid, while the mobile phase may be acetonitrile, glacial acetic or nitroethane containing various proportions of methanol, acetone or water. Other more exotic stationary phases that have been used but never widely adopted include rubber,[53, 203] Factice,[203, 279] polyethylene[162] and Teflon (Fluon).[27] As the solvents are constant in composition throughout the analysis, differential refractometers can be used to detect and estimate lipids eluting from the columns[203, 279] and when both phases are comparatively volatile liquids, moving wire liquid chromatography detectors or the quantification procedures described above for adsorption chromatography may be used.

Components emerge from liquid–liquid chromatography in much the same manner as from GLC columns and the efficiency of the former can be calculated from the elution profiles of the components in terms of numbers of theoretical plates using a relationship between their retention and elution volumes (c.f. retention time and width of peak) similar to that described earlier in this chapter for GLC columns. Maximum resolution should be obtained with narrow bore columns and small loads, but high pressure controlled-flow solvent delivery systems are necessary in this instance.

Thin-layer reverse phase systems, in which the non-polar stationary phase is held on a layer of inert support material on a glass plate and developed in a tank containing the polar mobile phase, have also been developed. Such plates are impregnated by slowly and very gently immersing plates, coated in the normal manner with a layer of silica gel or Kieselguhr, in a 5–10 per cent solution of the chosen stationary phase in a volatile solvent. When it is thoroughly soaked, the plate is removed and the volatile solvent is allowed to evaporate off in the atmosphere at room temperature. In a second method, the plate is developed in a tank containing a solution of the stationary phase then is removed and dried off as before. Finally, Litchfield[305] makes plates for reverse phase TLC by preparing a slurry consisting of hexadecane (4 g) in hexane (85 ml) and silanised silicic acid (50 g). After thorough shaking, the mixture is poured into a conventional TLC spreader and the layers are made in the normal way. The plates must be left overnight in a ventilated area to allow the hexane to evaporate before being used. If the stationary phase is a high boiling silicone such as Dow Corning 200 fluid, or a high molecular weight hydrocarbon such as tetradecane or hexadecane, the plates can be stored for long periods in sealed containers (undecane may evaporate off, however). Samples are applied to the plate in a volatile solvent such as pentane which is allowed to evaporate before the plate is placed in a tank containing the mobile phase 80–100 per cent saturated with stationary phase and is allowed to develop in the normal manner.

Theoretical aspects of reverse phase TLC have been discussed by Vereshagin.[555] The procedure has one important advantage over adsorption TLC in that Rf values are more reproducible. The main drawbacks are that it can be messy and that lipids are difficult to detect, especially in preparative applications, as the stationary phase interferes with non-specific non-destructive spray reagents. None the less, there is a report that 2′,7′-dichlorofluorescein sprays can be used to detect lipids on Kieselguhr impregnated with liquid paraffin;[581] lipids appear as green spots on a purple background under UV light. In purely analytical separations, it is possible to heat the plates to a sufficiently high temperature to evaporate off such stationary phases as undecane and then detect the lipids by charring techniques. Most organic compounds can also be detected on hydrophobic layers by iodine vapour or with a spray of 5 per cent phosphomolybdic acid in ethanol (followed by heating at 120°C for 5 min when lipids appear as blue spots) and chromic acid solution may be used on siliconised layers. All these procedures are destructive to lipids so are not suitable for preparative chromatography. For this purpose, it is usually necessary to compromise and destroy part of the sample by exposing one edge of the developed plate to iodine vapour to locate the ends of the bands, which can then be removed from adjacent regions of the plate. Lipids are recovered from these bands in the manner described for adsorption TLC but it may be necessary to subject the components isolated to adsorption chromatography to completely eliminate any adhering stationary phase.

Compounds that emerge as critical pairs on reverse phase chromatography can usually be resolved by chromatography on adsorbents impregnated with silver nitrate (see following section). Indeed, two-dimensional TLC systems in which components are separated in one direction by silver nitrate chromatography and in the second direction by a reverse phase system have been described.[270] Reverse phase systems have also been utilised to separate normal fatty acids from those containing polar functional groups, but adsorption chromatography is more convenient and is now generally preferred for the purpose.

5. *Chromatography on adsorbents containing complexing agents*

(i) *Silver nitrate chromatography*

In silver nitrate (or "argentation") chromatography, the property exhibited by silver compounds in that they form polar complexes reversibly with the double bonds of the aliphatic moieties of lipids is used as a means of separating them according to the number, configuration (*cis* or *trans*) and to some extent the position of those double bonds. The first procedure in which the principle was applied was in a countercurrent distribution system, but it is now more generally used in conjunction with adsorption chromatography. Morris[356] has reviewed the principle and applications of the method.

For silver nitrate thin-layer chromatography, 5 or 10 per cent by weight of silver nitrate relative to the weight of adsorbent is incorporated into the slurry used to make the plates. These are activated in the usual way and will retain their activity for a month or so if stored in the dark in a desiccator. On exposure to light, the plates blacken rapidly and it is preferable that they be handled and developed in a darkened room or cupboard. Lipids on the plate can be visualised under UV light after spraying with 2′,7′-dichlorofluorescein solution, when they appear as yellow spots on a red-purple background, and they can be recovered from the plates as described earlier for adsorption chromatography, although special precautions may be necessary to remove small amounts of silver that are also eluted. Ion exchange resins may be used to remove silver from phospholipids separated in this way,[295] but washing the extract in hexane:ether (1:1 v/v) with sodium chloride solution and then 0·05 M tris buffer (pH 9) or with dilute ammonia solution will remove both the silver and the dye from simple lipids.[15]

Silicic acid impregnated with silver nitrate has been used in columns to effect separations of lipids, and acid-washed Florisil so treated[23, 583] is particularly useful as it permits the separation of much larger quantities of lipid. Acid-washed Florisil, prepared as described earlier in this chapter, is impregnated with silver nitrate as follows (basically the procedure of Willner[583] which is more economic in its use of silver nitrate than others described).

"Acid-washed Florisil (35 g) and 14 ml of a 50 per cent aqueous solution of silver nitrate are mixed and the mixture shaken until free flowing (20 min). It is then left overnight in an oven at 120°C to activate, cooled and stored in the dark."

The adsorbent is packed into columns in the conventional manner except that the columns are wrapped in black paper or aluminium foil to exclude light. Silver complexes are also more stable at lower temperatures so cooling the columns with a water jacket may aid the separations. Improved silver nitrate–TLC separations at low temperatures have also been recorded. Ion exchange resins impregnated with silver nitrate have some advantages over adsorption columns as the silver is not eluted by polar solvents.[130, 602] Chloroform cannot be used as an eluting solvent for the columns as it reacts with silver nitrate with the formation of nitric oxide.

Silver nitrate chromatography has a singularly useful application in the separation of *cis-* from *trans-*olefinic compounds, a feat that is not easily attained by gas chromatography, for example. A further important advantage of the method is that traces of silver left with the lipids separated by this technique afford some protection against autoxidation.[496] In combination with gas chromatography or liquid partition chromatography, silver nitrate chromatography is a powerful method of isolating simpler molecular species of lipids.

(ii) *Glycol-complexing agents*

Chromatographic adsorbents impregnated with sodium borate or boric acid are useful for separating compounds with adjacent free hydroxyl groups. For example, vicinal diols of the *threo-* and *erythro-* configuration form non-polar complexes reversibly with borates and compounds containing these groups, which are not normally separable by adsorption chromatography, can be separated; the *threo*-complexes form more readily and are less polar than the *erythro*-compounds so have greater mobility on the impregnated adsorbents.[355] Partial glycerides can also be separated on boric acid impregnated adsorbents where the complexing agent, in addition to improving the separations, stabilises the compounds and prevents acyl migration.[535] Boric acid also forms complexes with ceramides having a double bond in position 4 of the long-chain base (although the reason for it is not known) and the property has been used in the separation of molecular species of ceramides (see Chapter 8). Boric acid eluting with the lipids can be removed by washing them in diethyl ether with ice-cold distilled water (the solution should be dried over anhydrous sodium sulphate immediately).

Boric acid impregnated layers on TLC plates are prepared by incorporating sufficient boric acid in the slurry used to prepare the plates so that it comprises 10 per cent by weight of the final adsorbent[355] and plates are activated and stored in the usual way. Acid-washed Florisil impregnated with boric acid for column chromatography can be prepared by adding to the adsorbent methanol containing sufficient boric acid to amount to 10 per cent by weight of the adsorbent, evaporating off the solvent on a rotary evaporator and activating at 110°C for 1 hr.[478]

Sodium arsenite impregnated layers also give remarkable separations of diols including diastereoisomers of these.[355] The free diols must be recovered by alkaline hydrolysis of the stable complex that is formed, after this is eluted from the adsorbent with solvents. The separations obtained with borate and arsenite impregnated layers differ in nature so the two systems may be used to complement each other.

C. Some Practical Considerations

1. *Autoxidation of lipids*

The problem of autoxidation of lipids has already been mentioned briefly in Chapter 2. It can give rise to very real difficulties in the chromatography of lipids and the greater the degree of unsaturation of the fatty acid components, the greater the risk. For example, linoleic acid is autoxidised twenty times as readily as oleic acid and each additional double bond in the molecule increases this rate by a factor of at least two.

To prevent autoxidation, lipids should be handled wherever possible in an atmosphere of nitrogen and antioxidants such as BHT[371, 600] (see Chapter 2) should be added at a level of 0·05 per cent to thin-layer chromatographic solvents and spray reagents in particular. Solvents should also be flushed with nitrogen to displace any dissolved oxygen. BHT volatilises in a stream of nitrogen on a boiling water bath and also is eluted with the solvent front in most TLC systems so can be removed from lipids easily if this is necessary for any purpose. It is more troublesome in gas chromatography as it chromatographs on many of the more useful polyester columns close to methyl myristate, which it may obscure, but if this is not a major fatty acid component, as in many animal or plant lipids, the nuisance value of the effect may be negligible. BHQ (1,4-dihydroxy-2-*tert*-butyl benzene) has been recommended as an alternative to BHT.[371] It has similar elution characteristics to BHT in adsorption chromatography but is eluted much later on GLC. On occasion, it may appear as little more than a slight elevation of the base line but in other circumstances it may obscure essential components of the sample analysed.

As mentioned in Chapter 2, large volumes of solvents should be removed from lipids in a rotary film evaporator at or near room temperature. Small volumes of solvent can be evaporated by directing a stream of nitrogen onto the surface of the solution. Purified lipids separated by chromatographic procedures should be stored in non-alcoholic solvents (such as chloroform) in sealed containers flushed with nitrogen at −20°C. If they are to be stored for long periods, it is necessary to seal them in glass vials under vacuum.

2. *Solvents*

All solvents, including high purity analytical grades, contain traces of impurities some of which, for example antioxidants, may have been deliberately introduced by manufacturers. As large volumes of solvents are often required for the isolation of very small amounts of lipid, serious contamination can occur. At the very least, solvents should be redistilled before use and on occasion more extensive purification may be necessary.

Diethyl ether and other ethers develop peroxides on storage and antioxidants are frequently added by the manufacturer to minimise this. Distillation from potassium hydroxide pellets is effective in removing most impurities (including much of the water) but a large volume should not be purified unless it is to be used quickly. Care must be taken to ensure that the distillation flasks are never allowed to boil dry as serious explosions can occur. Last traces of peroxide can be removed by percolating the solvent down a column of activated alumina. When dry ether is required for any purpose, sodium wire may be added to solvent freshly distilled from potassium hydroxide.

The chloroform sold by most manufacturers contains 0·25–2 per cent ethanol, which codistils with it, as a stabiliser. In this form, the solvent is much more polar and has quite different elution properties for lipids in adsorption chromatography than has the alcohol-free solvent. If necessary, the ethanol can be removed by washing the chloroform several times with half its volume of distilled water, drying the product over anhydrous sodium sulphate and storing in the dark (to minimise the photochemical formation of phosgene) over anhydrous granular calcium chloride. Alcohol-stabilised chloroform is used for most chromatographic purposes detailed later, however, unless a statement is made to the contrary.

Hexane and most aliphatic hydrocarbons may contain olefins that can be removed prior to distillation by washing with concentrated sulphuric acid followed by 10 per cent sulphuric acid containing potassium permanganate. Some manufacturers now supply hydrogenated solvents, however.

Benzene usually contains water, most of which distils over in the first small fraction that is discarded and it can be maintained in the dry state by storing in the presence of sodium wire. It is not always recognised that benzene is a highly toxic compound and great care must be taken so that there is little escape of the vapour into the laboratory atmosphere.

Methanol may contain traces of aldehydes and amines which can be largely removed by distillation from potassium hydroxide pellets. When dry methanol is required, it is prepared as follows:

"10 g of dry magnesium turnings and 30–50 ml of methanol are warmed together with a few crystals of iodine. A vigorous reaction soon starts and when all the magnesium has dissolved, a further 1·5 litres of methanol are added and the mixture is refluxed for 1 hr after which it is distilled. The first few mls should be discarded but most of the remainder may be collected."

All solvents, particularly ethers or those containing halogens, should be stored out of direct sunlight in brown bottles. More detailed purification procedures for solvents can be found in textbooks on laboratory practice in organic chemistry such as that by Vogel.[562]

3. *Contaminants*

Apart from the potential contaminants in solvents, extraneous lipid-like materials may appear from a variety of sources. Plasticisers (discussed earlier in Chapter 2) are the most common of these, but others may achieve significance when very small amounts of individual lipid components are separated for analysis. All laboratory reagents whatever their nature may from time to time contain troublesome impurities and vigilance must be exercised continuously to detect and eliminate these. For example, 2′,7′-dichlorofluorescein may contain impurities that give rise to spurious peaks on the GLC recorder trace unless the reagent is washed with hexane before use.[394] Other contaminants may arise from fingerprints, soaps, hair preparations, tobacco smoke, laboratory greases, the exhausts of vacuum pumps, floor polish and so forth; the list could be endless.

D. Spectroscopy

Infrared (IR) spectroscopy was the first of the spectroscopic methods to be applied to the analysis of lipids in general and fatty acids in particular and, although there have been comparatively few new developments involving the technique in recent years, it is probably still the spectroscopic method that would be chosen first for the preliminary examination of an unknown lipid. Ultraviolet (UV) spectroscopy is now used much less frequently by lipid analysts than formerly, but is still has important specific applications. On the other hand, in the last few years nuclear magnetic resonance (NMR) spectroscopy and mass spectrometry have been applied with great success to problems of lipid structure determination and new applications are continually being developed. IR, UV and NMR spectroscopy are non-destructive techniques so that samples may be recovered if necessary for further analysis; mass spectrometry requires so little material that it can often be considered in much the same light.

Most spectroscopic methods are based on empirical collations of vast amounts of data obtained from model compounds of known structure and in the interpretation of spectra of unknown compounds, a knowledge of these data is required. It will only be possible to reproduce a small part of such information here but further details can be obtained from more specialised publications. In particular, Chapman[83] has reviewed the application of spectroscopic techniques to lipid structural analysis in great detail. The general principles and range of applications of these techniques are now discussed; specific applications will be dealt with later in the appropriate sections.

1. *Infrared absorption spectroscopy*

Infrared spectra are obtained when energy of infrared light of a given frequency is absorbed by a molecule so that the amplitude of the vibrations of specific bonds between atoms in the molecule is increased. The most useful and conveniently measured region of the infrared spectrum (limited by sodium chloride optics) is over a range of wavelength of 2·5–15 μm (equivalent to wave numbers of 4000–667 cm^{-1}) although the *near* infrared region (0·8–2·5 μm or 12,500–4000 cm^{-1}) may also contain features of interest in lipid analysis. Bonds absorb energy and vibrate largely in two ways; *stretching* vibrations, in which the distance between atoms increases and decreases, and *bending* vibrations, in which the position of an atom is altered relative to the line of the original bond. Bonds between specific atoms in defined molecular environments absorb energy at certain fundamental frequencies that are characteristic of the particular bond and which can often be used to diagnose their presence in unknown molecules. These fundamental frequencies are affected only slightly by the presence of other atoms in the molecule, but minor shifts of frequency can in themselves sometimes be useful diagnostic aids.

With modern double-beam infrared spectrometers, the spectrum from 2·5 to 15 μm is scanned continuously and the percentage transmission through the sample is recorded directly on a chart graduated linearly in wave numbers (cm^{-1}) or more commonly in wavelengths (μm). Samples may be in the vapour phase, in solution, or as solids and liquids, and very fast scanning instruments are now available so that spectra can be obtained from compounds in the effluent of a gas chromatograph. Carbon tetrachloride and carbon disulphide are chosen most frequently as solvents as they are transparent over a wide range of wavelengths and can be used safely in cells with windows of rocksalt. Smaller samples or compounds that are insoluble in these solvents are analysed in potassium chloride or potassium bromide pellets. Samples and solvents must be thoroughly dried as water obscures part of the spectrum and may attack the cell windows, but further practical details can be found in the many excellent reviews and textbooks on the subject. Very small samples are best analysed as a smear in the centre of a sodium chloride disc in an instrument fitted with a beam condenser.

When infrared spectroscopy is used as a quantitative technique, the Beer–Lambert law is applied, i.e.

$$\log_{10}\frac{I_0}{I} = E = Kcl$$

where I_0 is the intensity of incident radiation, I the intensity of transmitted radiation, c the concentration (in gm/l), K the specific extinction coefficient, l the path length (in cm) and E the absorbance. The molecular extinction coefficient (ε) can be obtained from the relationship:

$$\varepsilon = \frac{E}{cl} \qquad (c \text{ is in gm moles/l})$$

Alternatively, when the molecular weight of a substance is not known, the intensity of absorption is defined by $E_{1\,cm}^{1\%}$, the absorbance of a 1 per cent solution of the substance in a 1 cm cell.

$$10\varepsilon = E_{1\,cm}^{1\%} \times \text{mol. wt.}$$

Normally the most prominent or characteristic absorption band of the compound to be estimated is selected, ideally one on a plateau so that a small change in the molecular environment of the absorbing group does not produce a significant change in the measured absorption, although it may produce a small change in the fundamental frequency. The linearity of the Beer–Lambert law must be checked with appropriate standards and if this holds, the concentration of the component to be measured can be obtained directly from its specific extinction coefficient. If the law does not hold, the concentration can be determined by reference to a calibration curve obtained with standards prior to the analysis. Such procedures are used to determine the *trans*-double bond content of lipids,[534] or to measure the amounts of various neutral lipids eluted from TLC plates,[141] for example. When two or more components are present in a mixture, the infrared absorption must be measured at two or more wavelengths. The curves of concentration plotted against absorbance are obtained from the individual pure compounds and, if E is measured at each wavelength and K for each compound is known, the concentration of each can be determined by solving simultaneous equations.

A great deal of valuable information about the physical states of lipids can be obtained from their spectra in the solid state but is outwith the scope of this text. Solution spectra are much simpler and much more useful for detecting and estimating specific functional groups in lipids. The frequencies due to the carbonyl function of most lipids are easily seen and identified, but *cis*-double bonds do not give rise to major absorption bands that lend themselves to characterisation in the usual infrared region (bands in the near infrared spectra can be used for this purpose, however). *Trans*-double bonds, on the other hand, do give rise to prominent absorption bands as do many other functional groups which may be found in the fatty acid chain. Organic bases of phospholipids also absorb at distinct and characteristic frequencies that can be used to aid their identification.[36, 370] The application of infrared spectroscopy to lipid analysis has been frequently reviewed.[82, 83, 140, 387]

2. *Ultraviolet absorption spectroscopy*

The ultraviolet spectrum of a compound is generally measured over the range 220–400 nm although the visible range up to 800 nm may also be covered by a single instrument. Quartz cells are transparent over this range and compounds are dissolved in a solvent such as 95 per cent ethanol for analysis. Conjugated double bonds absorb strongly in this region and before the development of gas chromatography, the concentrations of the main classes of polyunsaturated fatty acids were commonly determined by isomerising the double bonds with strong alkali at elevated temperatures, to form conjugated systems when the absorption at the characteristic wavelengths for conjugated dienes, trienes and so forth were determined using the Beer–Lambert equations described earlier for infrared spectroscopy.

Nowadays UV spectroscopy is used principally in the analysis of natural fatty acids containing conjugated double bond systems or for the estimation of conjugated dienes formed by the action of lipoxygenase or as a result of autoxidation. These and other uses of UV spectroscopy have been discussed in greater detail elsewhere.[83, 397]

3. *Nuclear magnetic resonance spectroscopy*

The nuclei of certain isotopes are continuously spinning with an angular momentum that can give rise to an associated magnetic field. If a very powerful external magnetic field is then applied to the nucleus and made to oscillate in the radio frequency range, the nucleus will resonate between different quantised energy levels at specific frequencies, absorbing some of the applied energy. Such very small changes in energy can be detected, amplified and displayed on a chart. The tracing obtained of the variation in the intensity of the resonance signal with increasing applied magnetic field is the NMR spectrum. In organic compounds, the isotope of hydrogen, ^{1}H, displays this phenomenon whereas the main isotopes of carbon, oxygen and nitrogen do not, so the resonance frequencies of hydrogen atoms in molecules are those most often measured and the technique in this instance is often referred to as proton magnetic resonance spectroscopy.

The frequency at which any given hydrogen atom in an organic compound resonates is strongly dependent on its precise molecular environment and is subjected to "chemical shifts" by the presence of other adjacent atoms. The fine structure of each absorption band is determined by the weak magnetic forces of neighbouring hydrogen atoms and this feature (known as spin–spin coupling), together with the extent of the chemical shift, may be used to deduce the arrangement of hydrogen atoms and often the complete structure of an unknown compound. As the area under an absorption band is proportional to the number of hydrogen atoms in an identical environment in the molecule responsible for that signal, additional information on the structures of unknown compounds can be obtained by integrating the signals while recording the spectrum. The theory of NMR spectroscopy is complex and the reader should consult one of the excellent textbooks on the subject before tackling the interpretation of spectra.

There are basically two main types of NMR spectroscopy, broad band and high resolution. The former can be used to obtain much valuable information on the physical state and environment of lipid molecules but has limited use as an analytical tool and will not be considered further in this work. High resolution NMR spectroscopy is a valued technique for obtaining information on the structures of unknown lipids and has been widely used to solve lipid structural problems. Instruments operating at 60 MHz are now widely available in chemical laboratories and more powerful instruments are coming into use; for example, NMR spectra of unsaturated fatty acids recorded at 220 MHz have recently been published.[142]

Compounds must be in solution for analysis and the solvent should preferably not contain the isotope ^{1}H. Carbon tetrachloride is suitable for non-polar lipids but deuterochloroform may also be used and deuterated methanol has been added to this to effect solution of phospholipids. Chemical shifts are not measured in absolute units but are recorded as parts per million of the resonance magnetic field. Tetramethylsilane is added to the solvents as an internal standard, and in the conventional system it is given the arbitrary value 10 on the so-called τ (tau) scale. (On the usual charts, values increase from left to right with the increasing strength of the magnetic field.) In a less frequently used system (the δ scale), tetramethylsilane is given the value zero. To convert:

$$\delta = 10 - \tau.$$

Approximately 50 mg of lipid is usually required for an analysis, more than for most other spectroscopic methods, but less material can be used if the instrument is coupled to a computer that can average the results of repeated scans of the spectrum and eliminate much of the background noise. All the material is of course recoverable for further analysis.

The technique has been widely used for lipid structure determinations, particularly for the identification and location of double bonds in fatty acids but also to detect and locate other functional groups such as cyclopropane rings, hydroxyl groups, triple bonds and oxirane rings in fatty acids. Glycerides, glyceryl ethers, glycolipids and phospholipids also have distinctive spectra which aid in their identification. The application of NMR spectroscopy in the analysis of lipids has been the subject of several detailed reviews.[83, 163, 223, 224]

4. *Mass spectrometry*

In the mass spectrometer, organic compounds in the vapour phase are bombarded with electrons and form positively charged ions that can fragment in a number of different ways to give smaller ionised entities. These ions are propelled through a powerful magnetic field and separated according to their mass to charge (m/e) ratio. Ions are collected in sequence as the ratio increases on a suitable detection system and are displayed as peaks on a chart. The largest peak or base peak in the spectrum is given an arbitrary intensity value of 100 and the intensities of all the other ions are normalised to this so that data can be presented in a uniform manner. The ion with the highest m/e value is generally (although not always) that of the original ionised molecule and is termed the parent ion (M^+). With low resolution instruments, peaks appear at unit mass numbers, but at higher resolutions the masses of individual ions can be measured with sufficient accuracy for the molecular formula of each to be unequivocally determined.

Molecules do not fragment in an arbitrary manner but tend to split at weaker bonds, such as those adjacent to specific functional groups, or according to certain complex rules which are now fairly well understood and have been formulated from the analysis of model compounds. The mass numbers and molecular formulae of the major ions can then be used to detect the presence and location of specific functional groups in a compound. Indeed, an accurate determination of the molecular weight and molecular formula of a compound can in itself be a valuable guide to its identity.

Mass spectrometry cannot be used to locate the position of double bonds in alkyl chains unless they are first hydroxylated and converted to non-polar derivatives that give definitive spectra; other functional groups, on the other hand, such as keto, hydroxyl, epoxyl and some cyclic groups in the fatty acid chains, are readily identified and their positions confirmed. In particular, methyl branches, which are not easily located by other methods, are easily seen by mass spectrometry. The technique has also been used for the identification of higher molecular weight lipids such as ceramides, long-chain bases and glycerides, but it is most useful for the analysis or identification of pure compounds rather than natural mixtures. It is rarely used as a quantitative technique although it has been used in this way to determine molecular species of triglycerides.[206] Mass spectra of model phospholipids have also been obtained, but so far the procedure has not been used as an aid to the identification of natural compounds of this type. It is possible to connect the outlet of a gas chromatograph through a suitable interface to a mass spectrometer so that individual compounds are identified as they are eluted; this has proved a particularly powerful tool in the hands of lipid analysts. It is advisable to use highly thermostable silicone phases in the GLC columns so that material bleeding from them does not contaminate the samples and/or the GLC–mass spectrometer interface. Although samples are destroyed during mass spectrometry, very small amounts are necessary for an analysis (<0.1 mg) so that this is not a serious disadvantage. The application of mass spectrometry to lipid structure determinations has recently been reviewed.[83, 319, 320]

5. *Optical rotary dispersion*

The determination of the absolute configuration of a compound is a problem not often faced by lipid analysts but nonetheless of considerable importance. Optical rotary dispersion measurements with modern recording spectropolarimeters are essential to this study, but the problem of assignment of configuration is too complex to lend itself to simple rationalisations. Fortunately, Smith has comprehensively reviewed problems of fatty acid and long-chain base asymmetry[502] and glyceride chirality.[503]

E. Equipping a Laboratory for Lipid Analysis

Most conventional chemical laboratories can be used for lipid analysis, but certain features are particularly desirable. For example, sealed bench tops that do not need waxing are useful as the waxes may be picked up on glassware and contaminate lipid samples. Indeed, sealed bench tops and floors are essential if isotopically-labelled lipids must be analysed. An efficient fume cupboard or hood is desirable for handling toxic reagents or isotopically-labelled lipids and for spraying TLC plates. As many of the solvents required for lipid analysis are highly inflammable, special precautions against accidental fires should be taken.

Conventional glassware with ground-glass joints can be used for chemical manipulations and too wide a range of joint sizes should be avoided for maximum interchangeability. As a great deal of lipid analytical work can be performed on a small scale, a range of semi-micro glassware is valuable. The author makes extensive use of test-tubes with ground-glass joints of about 15 ml capacity for carrying out chemical reactions. When it is necessary to extract aqueous solutions with organic solvents in such tubes, the layers are separated by means of Pasteur pipettes. The test-tubes can be used in standard centrifuges or can be equipped with small condensers when solvents must be heated under reflux. An aluminium heating block drilled with holes of the correct size is valuable in the latter circumstance. Rotary evaporators are essential for the removal of large volumes of solvent and these should have greaseless joints and be constructed so that solvent vapours are only permitted to come in contact with glass surfaces.

A supply of purified nitrogen should be available in the laboratory so that an inert atmosphere can be maintained in vessels containing lipids. If a stream of nitrogen is used to remove small volumes of solvents from lipid samples or to dry developed TLC plates, this should be performed in a fume cupboard.

D

The chromatographic equipment required will depend on the needs of the individual, but some general suggestions can be made. A gas chromatograph will probably be essential but should be situated in a separate room or as far as possible from benches where inflammable solvents may be handled. Those features of the instrument desirable for lipid analysis have been discussed above and are also considered in Chapters 5 and 8. For safety reasons, the cylinders containing the carrier and detector gases should be situated outside the laboratory. The thin-layer chromatography equipment required will probably comprise an adjustable spreader, plate leveller, glass plates of various sizes, developing tanks and airtight boxes or desiccators for storing activated plates. A range of glass columns will probably be required for column chromatography.

It is impossible to list all the apparatus that will be required as needs will vary with the nature of the work, but centrifuges, shakers, UV lights, magnetic stirrers, vortex mixers, heating mantles, thermostatically-controlled water baths and balances are likely to be needed from time to time and there should be access to spectrophotometers. Similarly, it is not possible to give a definitive list of chemicals that will be used, but a range of common organic solvents, thin-layer and column adsorbents, anhydrous sodium sulphate, potassium bicarbonate and sodium or potassium hydroxide will certainly be needed in most laboratories. A range of lipid standards is required and these together with lipid samples should be stored in a refrigerator or deep-freeze.

CHAPTER 4

The Preparation of Volatile Derivatives of Lipids

MANY simple and complex lipids are either too polar or of too high molecular weight to be subjected to some chromatographic procedures. Therefore, it is frequently necessary to convert them to volatile and/or non-polar derivatives for further analysis. In addition, it may be necessary to hydrolyse or transesterify lipids to obtain the component parts for a complete analysis of these. Because of the high sensitivity of thin-layer and gas chromatographic analysis procedures, very small amounts of material (1–50 mg) may be all that is required and most of the procedures described below are on this scale. Any conventional Pyrex glassware with ground-glass joints can be used for the reactions, but for many the author has found it convenient to use test-tubes of about 15 ml capacity with a ground-glass joint at the top. Organic and aqueous layers can be separated efficiently in these with the aid of Pasteur pipettes. Precautions should be taken to prevent autoxidation of lipids (see Chapter 3).

A. Saponification of Lipids

Lipids may be hydrolysed by heating them under reflux with an excess of dilute aqueous ethanolic alkali and the fatty acids, diethyl ether-soluble non-saponifiable materials and any water-soluble hydrolysis products recovered for further analysis. When the water-soluble components (such as glycerol, glyceryl phosphoryl choline, etc.) are required, special procedures must be used and these are discussed in later chapters. The free fatty acids and the diethyl ether-soluble non-saponifiable components are separately recovered in the following procedure:

"The lipid sample (1 g) is hydrolysed by refluxing it with a 1 N solution of potassium hydroxide in 95 per cent ethanol (6 ml) for 1 hr. The solution is cooled, water (12 ml) is added and the solution is extracted thoroughly with diethyl ether (3×10 ml). It may be necessary to centrifuge to break any emulsions that form. The extract is washed several times with water, dried over anhydrous sodium sulphate and the non-saponifiable materials are recovered on removal of the solvent in a rotary evaporator. The water washings are added to the aqueous layer which is acidified with 6 N hydrochloric acid and extracted with diethyl ether or hexane (3×10 ml). The free fatty acids are recovered after washing the extract with water, drying it over anhydrous sodium sulphate and removing the solvent in the usual way."

The non-saponifiable layer will contain any hydrocarbons, long-chain alcohols and sterols originally present in the lipid sample in the free or esterified form. If the sample contained any glyceryl ethers or plasmalogens, the deacylated residues will also be in this layer. When short-chain fatty acids are present in the lipids (C_{12} or less), it is necessary to extract the acidified solution much more exhaustively and even so it may be almost impossible to recover very short-chain fatty acids such as butyric quantitatively. Epoxyl groups and cyclo-propene rings in fatty acids are normally disrupted by acid, but with care they will survive the above procedure if the exposure to the acidic conditions is short.

Cholesteryl esters are hydrolysed very slowly by most reagents and may not have reacted completely if the above method is used, so if they are major components of the mixture, longer reflux times are necessary. Similarly, N-acyl derivatives of long-chain bases are not so readily saponified by alkali, but hydrolysis is virtually complete after about 16 hr at reflux.[251]

Non-saponifiable materials and free fatty acids can also be obtained by separating the total acidified extract by thin-layer chromatography using the solvent systems suggested in Chapter 6, eliminating the step in which the alkaline solution is extracted; the free fatty acids are easily separated from the other products of hydrolysis which can be individu-ally isolated and identified. As an alternative, acidic and neutral materials can be separated by ion exchange chromatography using the following procedure.[612]

"One gram of DEAE-Sephadex (Pharmacia Fine Chemicals, Sweden), type A-25 (capacity 3·2 m-equiv per g), is successively washed on a buchner funnel with small amounts of 1 N hydrochloric acid, water, 1 N potassium hydroxide and water again (the procedure is repeated three times), the last wash until neutral. It is then washed twice with methanol (25 ml) and with 25 ml of diethyl ether–methanol–water (89:10:1 by volume). It is slurried in the latter solvent mixture, equilibrated overnight and packed into a small column. The sample, containing no more than 100 mg of fatty acids, is washed through with 25 ml of the solvent mixture; the neutral materials are eluted while the acids remain on the column. The latter can be eluted with diethyl ether–methanol (9:1, v/v) saturated with carbon dioxide."

Polyunsaturated fatty acids are not altered by the mild hydrolysis conditions described above,[240] but if the reaction time is prolonged unduly or if too strong alkali is used, some isomerisation of double bonds may occur.

B. The Preparation of Methyl Esters of Fatty Acids

Before the fatty acid composition of a lipid is determined by gas chromatography, it is necessary to prepare the comparatively volatile methyl ester derivatives of the fatty acid components. This must be by far the commonest chemical reaction performed by lipid analysts yet it is often poorly understood; the topic has recently been reviewed by Christie.[91] There is no need to hydrolyse lipids to obtain the free fatty acids before preparing the esters as most lipids can be transesterified directly. No single reagent will suffice for all purposes, however, and one must be chosen that best fits the circumstances. Esters prepared by any of the following methods can be purified if necessary by preparative TLC on silica gel G layers with hexane–diethyl ether (9:1, v/v) as developing solvent.

1. *Acid-catalysed esterification and transesterification*

Free fatty acids are esterified and O-acyl lipids transesterified by heating them with a large excess of anhydrous methanol in the presence of an acidic catalyst. If water is present, it may prevent the reaction going to completion. The commonest and mildest reagent used for the purpose is 5 per cent (w/v) anhydrous hydrogen chloride in methanol. It can be prepared by bubbling hydrogen chloride gas (which is commercially available in cylinders or can be prepared by dropping concentrated sulphuric acid slowly on to fused ammonium chloride or into concentrated hydrochloric acid[562]) into dry methanol, but a simpler procedure is to add acetyl chloride (5 ml) slowly to cooled dry methanol (50 ml).[26] Methyl acetate is formed as a by-product but does

$$CH_3OH + CH_3.CO.Cl \rightarrow CH_3.CO_2CH_3 + HCl$$

not interfere seriously with the reaction at this concentration. It is usual to heat the reagent with the lipid sample under reflux for about 2 hrs, but they may also be heated together in a sealed tube at higher temperatures for a shorter period. Non-polar lipids such as cholesteryl esters or triglycerides are not soluble in the reagent and will not react in a reasonable time unless a solvent is added to effect solution. Benzene is most often used for the purpose, but dichloromethane is less toxic and equally effective unless phospholipids are present (partially hydrolysed phospholipids may precipitate from solution in this solvent).

A solution of 1–2 per cent (v/v) concentrated sulphuric acid in methanol transesterifies lipids in the same manner and at much the same rate as methanolic hydrogen chloride and it is very easy to prepare, but if the reagent is used carelessly, some decomposition of polyunsaturated fatty acids may occur. Boron trifluoride in methanol (12–14 per cent w/v) is also used as a transesterification catalyst and in particular as a rapid esterifying reagent for free fatty acids.[366] The reagent has a limited shelf life unless refrigerated and the use of old or too concentrated solutions may result in the production of artefacts and in the loss of large amounts of polyunsaturated fatty acids.[143, 312] In view of the large amount of acid catalyst used in comparison to other reagents and the many known side reactions, it is the author's opinion the boron trifluoride in methanol has been greatly overrated as a transesterifying agent although it is undoubtedly valuable for esterifying free fatty acids.

Methanolic hydrogen chloride (5 per cent) is then probably the best general purpose esterifying agent. It methylates free fatty acids very rapidly and can be used to transesterify other O-acyl lipids; it is generally used as follows:

"The lipid sample (up to 50 mg) is dissolved in benzene (1 ml) in a test tube and 5 per cent methanolic hydrogen chloride (2 ml) is added. The mixture is refluxed for 2 hr then water (5 ml) containing sodium chloride (5 per cent) is added and the required esters are extracted with hexane (2 × 5 ml) using washable or disposable Pasteur pipettes to separate the layers. The hexane layer is washed with water (4 ml) containing potassium bicarbonate (2 per cent) and dried over anhydrous sodium sulphate. The solution is filtered and the solvent removed under reduced pressure in a rotary film evaporator or in a stream of nitrogen."

If no phospholipids are present in the sample, dichloromethane can be substituted for benzene. No solvent is necessary if free fatty acids alone are to be methylated (also only 20 min at reflux is required) or if phospholipids are to be transesterified. The reaction can be scaled up considerably, although for very large amounts methanol containing concentrated sulphuric acid is probably more economical; for example, 50 g of lipid in benzene (100 ml) can be transesterified with 200 ml of methanol containing 4 ml concentrated sulphuric acid.

N-acyl lipids are transesterified very slowly with these reagents (see below). If acidic reagents are permitted to superheat in air, some artefact formation is possible.

2.   *Base-catalysed transesterification*

O-Acyl lipids are transesterified very rapidly in anhydrous methanol in the presence of a basic catalyst. Free fatty acids are *not* esterified, however, and care must be taken to exclude water from the reaction medium to prevent their formation as a result of hydrolysis of lipids. 0·5 N sodium methoxide in anhydrous methanol, prepared simply by dissolving fresh clean sodium in dry methanol, is the most popular reagent but potassium methoxide or hydroxide are also used as catalysts. The reagent is stable for some months at room temperature, especially if oxygen-free methanol is used in its preparation. The reaction is very rapid; phosphoglycerides, for example, are completely transesterified in a few minutes at room temperature. It is performed as follows:

> "The lipid sample (up to 50 mg) is dissolved in benzene (1 ml) in a test-tube and 0·5 N sodium methoxide in anhydrous methanol (2 ml) is added. The solution is maintained at 50°C for 10 min then glacial acetic acid (0·1 ml) is added followed by water (5 ml) and the required esters are extracted with hexane (2 × 5 ml), using a Pasteur pipette to separate the layers. The hexane layer is dried over anhydrous sodium sulphate containing 10 per cent solid potassium bicarbonate and filtered before the solvent is removed under reduced pressure on a rotary film evaporator."

As with acid-catalysed procedures, benzene is necessary to effect solution of non-polar lipids such as cholesteryl esters or triglycerides and is not required if they are not present in the sample. Dichloromethane can be substituted for benzene if no phospholipids are present, but chloroform cannot be used as dichlorocarbene, which can react with double bonds, is generated by reaction with the sodium methoxide. Cholesteryl esters are transesterified very slowly and may require twice as long a reaction time as that quoted. The quantities of lipid used can be scaled up considerably; for example, 50 g of lipid is transesterified in benzene (50 ml) and methanol (100 ml) containing fresh sodium (0·5 g) in 10 min at reflux, and related procedures have been used to transesterify litre quantities of oils.[368] Under the conditions described above no isomerisation of double bonds in polyunsaturated fatty acids occurs, though prolonged or careless use of basic reagents may cause alterations to fatty acids.

Amide-bound fatty acids, as in sphingolipids, are not affected by alkaline transesterification reagents under such mild conditions and this fact is sometimes used in the purification of these lipids. Also, aldehydes are not liberated from plasmalogens with basic reagents but this does occur under acidic conditions.

3. *Diazomethane*

Diazomethane reacts rapidly with unesterified fatty acids forming methyl esters in the presence of a little methanol which catalyses the reaction. The reagent is generally prepared in ethereal solution by the action of alkali on a nitrosamide, e.g. N-methyl-N-nitroso-*p*-toluene-sulphonamide ("Diazald", Aldrich Chemical Co., Milwaukee, U.S.A.) in the presence of an alcohol. Solutions of diazomethane are stable for short periods if stored refrigerated in the dark over potassium hydroxide pellets, but if kept too long polymeric by products form that may interfere with subsequent gas chromatographic analysis. Diazomethane is highly toxic and potentially explosive so great care must be exercised in its preparation, in particular strong light and apparatus with ground glass joints must be avoided.

The procedure of Schlenk and Gellerman[461] is particularly convenient for the preparation of small quantities of diazomethane for immediate use. In this instance there is very little by-product formation and, if sensible precautions are taken, the risk to health is negligible.

"A simple apparatus is required that can be quickly assembled by a glassblower. It consists of three tubes with side arms that are bent downwards and arranged so that the arm of each projects into and is near the bottom of the next tube. A stream of nitrogen is saturated with diethyl ether in the first tube and carries diazomethane, generated in the second tube, into the third tube where it esterifies the acids. The flow of nitrogen through diethyl ether in tube 1 is adjusted to 6 ml per min. Tube 2 contains carbitol (0·7 ml), diethyl ether (0·7 ml) and 1 ml of an aqueous solution of potassium hydroxide (600 g/l). The fatty acids (5–30 mg) are dissolved in diethyl ether–methanol (2 ml; 9:1 by vol) in tube 3. About 2 mmole of N-methyl-N-nitroso-*p*-toluenesulphonamide per mmole of fatty acid in ether (1 ml) is added to tube 2 and the diazomethane which is formed is passed into tube 3 until the yellow colour persists. Excess reagent is then removed in a stream of nitrogen."

When large amounts of diazomethane are needed, the procedure of De Boer and Backer[114] can be recommended.

4. *Special cases*

(i) *Short-chain fatty acids*

Short-chain acids are completely esterified by any of the procedures described above, but quantitative recovery of the esters from the reaction media is very difficult because of their high volatility and ready solubility in water. As such acids are major components of commercially important fats and oils such as milk fats or coconut oil, a great deal of attention has been given to the problem. Diazomethane can be used to esterify free fatty acids quantitatively in ethereal solution and the reaction medium injected directly into the gas chromatograph so that there are no losses, but if the free acids have to be obtained by hydrolysis of lipids, it is not easy to ensure that there are no losses at this stage. The best methods are those in which there are no aqueous extraction or solvent removal steps and in which the reagents are not heated. The alkaline transesterification procedure of Christopherson and Glass,[101] on which the following procedure is based, best meets these criteria.

"The oil (20 mg) is dissolved in hexane (0·3 ml) in a stoppered test-tube and 2 N sodium methoxide (0·1 ml) is added. The mixture is shaken gently for 5 min at room temperature then more hexane (2·5 ml) is added followed by powdered anhydrous calcium chloride. The mixture is allowed to stand for 1 hour then centrifuged at 2000–3000 rpm for 2–3 min to precipitate the drying agent. An aliquot of the supernatant liquid is taken for GLC analysis."

If the sample contains both O-acyl bound and unesterified fatty acids, the latter may be esterified with diazomethane first before the former are transesterified.

(ii) *Unusual fatty acids*

The methods described above can be used to esterify all fatty acids of animal origin without causing any alteration to them. Many fatty acids from plant sources and certain of bacterial origin, however, are more susceptible to chemical attack. For example, cyclopropane, cyclopropene and epoxy fatty acids are disrupted by acidic conditions and lipid samples containing these acids should be transesterified with basic reagents; the free fatty acids can be methylated safely with diazomethane. Conjugated polyenoic fatty acids such as α-eleostearic acid (see Chapter 1) undergo *cis–trans*-isomerisation and double bond migration when esterified with methanolic hydrogen chloride[277] and all acidic reagents can cause addition of methanol to conjugated double bond systems,[281] but no side effects occur when basic transesterification is used. Similar reactions occur under acidic conditions with fatty acids containing a hydroxyl group immediately adjacent to a conjugated double bond system (e.g. dimorphecolic acid, 9-hydroxy, 10-*trans*, 12-*trans*-octadecadienoic acid) and dehydration and other unwanted side reactions may also take place.

(iii) *Amide-bound fatty acids*

Sphingolipids, which contain fatty acids linked by N-acyl bonds, are not easily transesterified under acidic or basic conditions. If the fatty acids alone are required for analysis, the lipids may be refluxed with methanol containing concentrated hydrochloric acid (5:1 v/v) for 5 hr[262] or by maintaining the reagents at 50°C for 24 hr[91] and the products worked up as described above for the anhydrous reagent. Traces of degradation products of the bases that might interfere with subsequent analyses can be removed by TLC. If the long-chain bases are also required for analysis, sphingomyelin, for example, should first be dephosphorylated with phospholipase C (see Chapter 9)[251] before being hydrolysed by heating under reflux for 16 hr with 1 M methanolic potassium hydroxide as described above (section A). The organic bases are recovered in the "non-saponifiable" layer. Further details of such procedures are given in Chapter 7. With N-acyl phosphatidyl serine and related lipids, the O-acyl bound fatty acids can be released by mild alkaline methanolysis and so distinguished from the N-acyl components which require much more vigorous hydrolytic conditions.

(iv) *Esterification on TLC adsorbents*

After lipids have been separated by thin-layer chromatography, the conventional procedure is to elute them from the adsorbent before transesterifying for GLC analysis. However, it is possible to esterify lipids directly on the adsorbent, either by spraying the plate with an esterifying reagent (messy and not easily controlled) or better by scraping the adsorbent bands into a test-tube, adding the reagent directly to this (as silica gel will neutralise the sodium methoxide it is necessary to use a more concentrated reagent, i.e. 2 M) mixing thoroughly, and carrying out the reaction as when no adsorbent is present. On working up the aqueous mixture obtained when the reaction is stopped, it is necessary to centrifuge to precipitate all the silica gel and then to extract with a more polar solvent than hexane, for example diethyl ether, to ensure quantitative recovery of the methyl esters.[90] Unfortunately, cholesteryl esters are not transesterified readily in the presence of silica gel and it is still necessary to elute these from the adsorbent first. The elimination of the conventional elution step reduces the risk of loss of the more polar lipids and their fatty acid components and of contamination of the sample by traces of impurity in large volumes of solvent.

C. Derivatives of Free Hydroxyl Groups

The free hydroxyl groups of long-chain alcohols, hydroxy fatty acids and partial glycerides or glyceryl ethers are frequently converted to non-polar derivatives for chromatographic analysis. Acyl migration of fatty acids in partial glycerides is prevented and more symmetrical peaks are obtained on gas chromatography than could be obtained with the original compounds. The choice of derivative will depend on the nature of the compound and the separation to be attempted and on occasion it may be necessary to prepare several types of derivative to confirm identifications of components.

1. *Acetylation*

Acetyl chloride and pyridine at room temperature or prolonged heating with acetic anhydride can be used to acetylate lipids, but the mildest reagent is probably acetic anhydride in pyridine (5:1 v/v), which is used as follows.[428]

"The lipid (up to 50 mg) is dissolved in acetic anhydride in pyridine (2 ml, 5:1, v/v) and left at room temperature overnight. The reagents are then removed in a stream of nitrogen with gentle warming and the acetylated lipid is purified, if necessary, by TLC on silica gel layers. A solvent system of hexane–diethyl ether (80:20 v/v) is suitable for most acetylated simple lipids."

Free amino groups are also acetylated with this reagent; a procedure for acetylating amino groups without acetylating hydroxyl groups at the same time is described in Chapter 7.

Acetylated diglycerides can be prepared directly from phospholipids by heating them at 140°C with acetic anhydride and acetic acid in a sealed tube,[425] but considerable acyl migration occurs under these conditions [382,427] and it is preferable to hydrolyse phospholipids to diglycerides enzymatically with phospholipase C before acetylation (see Chapter 9).

2. *Trifluoroacetates*

Trifluoroacetate derivatives of hydroxy acids, monoglycerides and glyceryl ethers are extremely volatile and are sufficiently temperature stable to be subjected to gas chromatographic analysis. They are prepared by dissolving the hydroxy compound in excess of trifluoroacetic anhydride, leaving for 30 min and then removing the excess reagent on a rotary film evaporator.[594] Diglyceride trifluoroacetates are not sufficiently stable at high temperatures for gas chromatography, however.[285] Trifluoroacetates hydrolyse very rapidly even in inert solvents such as hexane or benzene and it is necessary to store them and to inject them onto the gas chromatographic column in trifluoroacetic anhydride solution. Column packings must be conditioned by repeatedly injecting trifluoroacetic anhydride into them before being used, but are rendered acidic and may no longer be suitable for other analyses. Trifluoroacetic acid also appears to attack the methylene group between double bonds causing losses of polyunsaturated components.[594]

3. *Trimethylsilyl ethers*

Trimethylsilyl (often abbreviated to TMS) ether derivatives are a useful alternative to acetates for gas chromatographic analysis. They are considerably more volatile than acetates but are not as stable, particularly to acidic conditions, and will hydrolyse on TLC adsorbents. The most popular reagent consists of a mixture of hexamethyldisilazane, trimethylchlorosilane and pyridine (3:1:10 by volume) and is used as follows.[593]

"To the hydroxy-compound (10 mg) is added pyridine (1 ml), hexamethyldisilazane (0·3 ml) and trimethylchlorosilane (0·1 ml). The mixture is shaken for 30 sec and allowed to stand for 5 min when hexane (5 ml) and water (5 ml) are added and the whole shaken. The hexane layer is removed and the aqueous layer washed twice more with 5-ml portions of hexane. The combined hexane layers are dried over anhydrous sodium sulphate then the solvent is removed in a stream of nitrogen until the odour of pyridine is gone. The derivatives need no further purification and are stable for long periods at −20°C."

A mixture of hexamethyldisilazane, pyridine and trifluoroacetic acid (9:10:1 by vol)[106] is used in a similar manner. A number of other reagents have been used and may be valuable for sterically hindered hydroxyl groups; they include N-trimethylsilylacetamide,[47] bis (trimethylsilyl) acetamide[276] and N-trimethylsilyldiethylamine.[504]

Dimethylsilyl ethers are more volatile than trimethylsilyl ethers but are much less stable than the latter unless very great care is taken in their preparation.[524]

4. *Isopropylidene compounds*

Isopropylidene derivatives of vicinal diols, for example glyceryl ethers or dihydroxy acids, are prepared by reacting the diol with acetone in the presence of a small amount of an acidic catalyst (see Fig. 4.1). Anhydrous copper sulphate is probably the mildest catalyst and is used as follows:

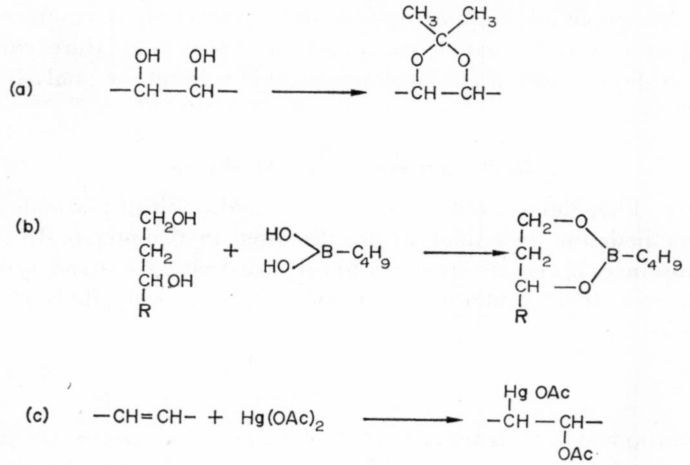

FIG. 4.1. Preparation of lipid derivatives. a. Isopropylidene derivatives. b. *n*-Butylboronate derivatives. c. Mercuric acetate adducts.

"The esters (10 mg) are dissolved in dry acetone (3 ml) and anhydrous copper sulphate (50 mg) is added. After 24 hr at room temperature (or 3 hr at 50°C), the solution is filtered, the copper salts are washed with dry ether and the combined solutions are evaporated *in vacuo*."

Alternatively, perchloric acid may be used as catalyst.[187] Isopropylidene compounds are stable under basic conditions but are hydrolysed by aqueous acid and the original diol compound can be regenerated by shaking with a solution of concentrated hydrochloric acid in 90 per cent methanol (1 M).

5. *n-Butylboronate derivatives*

Alkyl-boronic acids, such as n-butylboronic acid, react with 1,2- or 1,3-diols or α- or β-hydroxy acids to form 5- or 6-membered ring nonpolar boronate derivatives as illustrated in Fig. 4.1b. They are prepared simply by adding n-butylboronic acid to a solution of the hydroxy-compound in dimethylformamide; reaction is complete in 10–20 min at room temperature and the reaction mixture can be injected directly into a gas chromatographic column for analysis.[25, 67]

D. Derivatives of Fatty Aldehydes

Fatty aldehydes are obtained on acidic hydrolysis of plasmalogens, and methods for their analysis are discussed in Chapter 6. In some circumstances it may be necessary to prepare derivatives to adequately characterise them and the preparation of such compounds is now described.

1. *Hydrazone derivatives*

p-Nitrophenylhydrazones or 2,4-dinitrophenylhydrazones are particularly useful derivatives of aldehydes as they can be determined spectrophotometrically. They are prepared from the aldehydes or directly from the natural plasmalogens by the following procedure.[421]

"To a solution of the phospholipid (1–5 mg) in 95 per cent ethanol (1·6 ml) is added freshly prepared 0·02 M *p*-nitrophenylhydrazine in 95 per cent ethanol (0·2 ml), followed by 1 N sulphuric acid (0·2 ml). The solution is heated at 70°C for 20 min then cooled and water (1 ml) and hexane (2 ml) are added After shaking, the aqueous layer is removed and the hexane layer washed twice with water (2 ml) after which it is dried over anhydrous sodium sulphate and the solvent removed *in vacuo*."

Dimethylhydrazones are useful non-polar volatile derivatives of aldehydes as they can be readily separated by gas chromatography. They are prepared simply by dissolving the aldehyde in N,N-dimethylhydrazine and leaving at room temperature for 2 hr before the excess reagent is removed in a stream of nitrogen.[247]

2. *Acetals*

Dimethyl acetals are the simplest acetal derivatives of aldehydes and are prepared by heating the aldehydes under reflux with 5 per cent methanolic hydrogen chloride in the same manner as was described, earlier for the preparation of methyl esters. They can also be prepared directly from plasmalogens, but then have to be separated from methyl esters, which are formed at the same time, by adsorption chromatography or better by saponification of the esters.

Cyclic acetals or 1,2-dioxolanes, prepared by condensing 1,3-propanediol with aldehydes in the presence of an acidic catalyst, have greater thermal stability and are sometimes favoured for gas chromatographic analysis. They are prepared as follows.[522]

"The aldehydes (10 mg) with *p*-toluenesulphonic acid (0·5 mg) and 1,3-propanediol (50 μl) in 5 ml of chloroform are heated in a sealed tube at 80°C for 2 hr. On cooling, chloroform (3 ml), methanol (4 ml) and water (3 ml) are added and shaken. The lower layer, which contains the required derivatives, is taken to dryness."

Hydrazone derivatives of aldehydes can be converted directly to acetals by heating them with the required alcohol and an acid catalyst in the presence of acetone, which serves as an exchanger, by the following method.[330]

"The dinitrophenyl hydrazone (50 mg) is mixed with 10 per cent methanolic hydrogen chloride (5 ml) and acetone (5 ml) and refluxed for 45 min. The mixture is cooled to 0°C, neutralised with 5 N sodium hydroxide in 90 per cent methanol (5 ml) and extracted twice with hexane (50 ml portions). The hexane layer is washed with water, dried over anhydrous sodium sulphate and decolourised with charcoal before the required derivatives are recovered."

Cyclic acetal derivatives are prepared similarly.[554] All acetal derivatives are stable to alkaline conditions, but are hydrolysed by aqueous acid.

E. Derivatives of Double Bonds

The double bonds of unsaturated fatty acids may be reacted with formation of various addition compounds as an aid to the isolation of individual fatty acids or as a method of establishing the configuration or location of the double bonds in the aliphatic chain.

1. *Mercuric acetate derivatives*

Mercuric acetate reacts with the double bonds of unsaturated fatty acids to form polar derivatives (see Fig. 4.1c) that are much more easily separated by adsorption or partition chromatography than the parent fatty acids. They can be prepared by the following procedure.[515]

"The lipid sample is refluxed with a 20 per cent excess of the theoretical amount of mercuric acetate in methanol (2 ml per g of mercuric acetate) for 30 min. After cooling to room temperature, a volume of diethyl ether 2.6-fold that of the methanol is added to precipitate mercury salts and the solution filtered. The mercuric acetate adducts are obtained on removal of the solvent."

Although silver nitrate chromatography has supplanted the chromatography of mercuric acetate derivatives for most analytical purposes, such adducts may still be useful for the bulk perparation of pure fatty acids (see Chapter 5) and for certain synthetic applications. The original double bond is regenerated by reaction with aqueous acid with no double bond migration or *cis–trans* isomerisation if the following method is used:[515]

"10 g of the adduct in methanol (20 ml) is mixed with concentrated hydrochloric acid (50 ml) and a stream of hydrogen chloride gas is bubbled into the solution for a few minutes. The solution is extracted twice with hexane (50-ml portions) and the extracts are combined and washed with water. If a test of the organic layer with a solution of diphenylcarbazone in methanol reveals that mercury is still present, a single repeat of the process will remove it entirely."

Acetylenic groups react in a similar manner with 2 mols of mercuric acetate but cannot be regenerated as on acidic hydrolysis a keto derivative is formed which may be used, particularly in conjunction with mass spectrometry, to locate the position of the original triple bond.[71]

2. *Hydroxylation*

Double bonds can be oxidised to vicinal diols by a variety of reagents of which the most useful are alkaline potassium permanganate and osmium tetroxide. With both these reagents, *cis*-addition occurs to yield with *cis*-double bonds diols of the *erythro* configuration and with *trans*-double bonds diols of the *threo* configuration. The following procedure utilises alkaline permanganate and is particularly suited to reaction with monoenes and dienes.[380]

"The fatty acid ($0 \cdot 1$–$100 \, \mu$mole) is dissolved in $0 \cdot 25$ N sodium hydroxide solution ($0 \cdot 2$ ml) and diluted with ice water (1 ml). $0 \cdot 05$ M potassium permanganate ($0 \cdot 2$ ml) is added and after 5 min the solution is decolorised by bubbling sulphur dioxide into it. The fatty acid derivatives are extracted with chloroform–methanol (7 ml, 2:1 v/v) and the lower layer is dried over anhydrous sodium sulphate before the solvent is removed".

Osmium tetroxide gives higher yields of multiple diols from poly-unsaturated fatty acids.[380]

"The fatty acid ($0 \cdot 1$ to $100 \, \mu$ moles) is dissolved in dioxane–pyridine (1 ml, 8:1 v/v) and a 5 per cent solution of osmium tetroxide in dioxane ($0 \cdot 1$ ml) is added. After 1 hr at room temperature, methanol ($2 \cdot 5$ ml) and 16 per cent sodium sulphite in water ($8 \cdot 5$ ml) are added and the mixture allowed to stand for 1 hr more. After centrifuging to remove sodium sulphite, the supernatant solution is diluted with 4 volumes of methanol and filtered. The filtrate is evaporated to dryness and suspended in methanol (2 ml). Chloroform (4 ml) is added, the suspension is filtered and the solvent evaporated."

Such compounds in the form of less polar derivatives, for example isopropylidene[322] or methoxy[380] compounds, are used in conjunction with mass spectrometry to locate the position of the original double bond in the fatty acids (see Chapter 5). As the isopropylidene derivatives of *threo* and *erythro* compounds are separable by gas chromatography, the configuration of the original double bond can also be established.[322]

3. *Epoxidation*

Epoxides are formed from olefins by the action of certain per-acids. *Cis*-addition occurs and *cis*-epoxides are formed from *cis*-olefins and *trans*-epoxides from *trans*-olefins. The following procedure is based on that of Gunstone and Jacobsberg.[166]

"The monoenoic ester (20 mg) is reacted with *m*-chloroperbenzoic acid (16 mg) in chloroform (2 ml) at room temperature for 4 hr. Potassium bicarbonate solution (5 per cent, 4 ml) is added and the product is extracted thoroughly with diethyl ether. After drying the organic layer over anhydrous sodium sulphate and removing the solvent, the required epoxy ester is obtained by preparative TLC on silica gel G layers with hexane–diethyl ether (4:1 v/v) as developing solvent."

Peracetic or perbenzoic acids are also used in a similar manner for the purpose. Mass spectrometry of epoxy derivatives is sometimes used to locate the position of double bonds in fatty acids,[34,273] and as *cis*- and *trans*-isomers can be separated by GLC as a means of estimating fatty acids with *trans*-double bonds.[127,129]

4. *Deuterohydrazine reduction*

Deuterium can be added across double bonds to assist in their location by mass spectrometry by means of deuterohydrazine reduction.[117]

"The unsaturated ester (0·5 mmole) is dissolved in anhydrous dioxane (5 ml) at room temperature and deuterohydrazine (5 mmole) in twice its volume of deuterium oxide is added. The mixture is stirred at 55–60°C in air but in the absence of atmospheric moisture for 8 hr when the reagents are evaporated *in vacuo*. The product is purified by preparative silver nitrate TLC (see Chapter 5 for practical details.")

CHAPTER 5

The Analysis of Fatty Acids

A. Introduction

Gas–liquid chromatography is an extremely powerful tool for the determination of the fatty acid composition of lipid samples, but it must be strongly emphasised that fatty acids can only be tentatively identified by this technique and that complete confirmation of their structures must be obtained by unequivocal chemical methods. Ideally, individual pure fatty acids (usually in the form of the methyl ester derivatives, prepared as described in Chapter 4) should be isolated by chromatographic procedures and examined first by non-destructive spectroscopic techniques before chemical degradative procedures are applied. For example, adsorption chromatography will separate normal fatty acids from those containing polar functional groups. Unsaturated fatty acids of animal origin may be provisionally identified by their gas chromatographic retention characteristics on two or more stationary phases and on the basis of possible biosynthetic relationships. Silver nitrate chromatography can be used to segregate fatty acids according to the number and geometrical configurations of their double bonds; a portion of each fraction should be hydrogenated so that the lengths of the carbon chains of the components can be confirmed. Finally some form of partition chromatography must be utilised to separate components of different chain lengths so that the position and configuration of the double bonds may be determined by spectroscopic (principally IR and NMR spectroscopy) and oxidative degradation procedures. Appropriate spectroscopic and chemical techniques must also be used to detect and locate any other functional groups. Precautions should always be taken to minimise autoxidation (see Chapter 3).

103

B. Analytical Gas–Liquid Chromatography

1. *Fatty acids of animal origin*

(i) *Column packing materials*

The liquid phase used in preparing gas chromatographic column packing materials is the principal factor determining the nature of the separations that can be achieved. Silicone liquid phases such as SE-30, OV-1, JXR or QF-1 permit the separation of fatty acid esters mainly on the basis of their molecular weights, although there is separation of unsaturated from saturated fatty acids of the same chain length when the amount of the stationary phase on the support is low (1–3 per cent). High molecular weight hydrocarbon liquid phases such as the Apiezon greases (of which the most popular is Apiezon L) also separate saturated and unsaturated components of the same chain length, unsaturated esters eluting before the related saturated compounds, but with packed columns there is very little separation of esters of a given chain length differing in the number of double bonds in the molecule.

Polar polyester liquid phases are much more suited to fatty acid analysis as they allow clear separations of esters of the same chain length, but with zero to six double bonds, unsaturated components eluting after the related saturated compound. These phases can be subdivided into three classes: group a, highly polar phases, e.g. polymeric ethylene-glycol succinate (EGS), diethyleneglycol succinate (DEGS) and EGSS-X (a copolymer of EGS with a methyl silicone); group b, medium polarity phases, e.g. polyethyleneglycol adipate (PEGA), butanediol succinate (BDS) and EGSS-Y (a copolymer of EGS with a higher proportion of the methylsilicone than in EGSS-X); group c, low polarity phases, e.g. neopentylglycol succinate (NPGS) and EGSP-Z (a copolymer of EGS and a phenyl silicone). EGSS-X and EGSS-Y have now become widely accepted as the most useful representatives of the first two groups because of their relatively high thermal stability, particularly in packed columns, and it is to be hoped that more and more research workers will settle on these as standard phases for reporting gas chromatographic retention data. Low polarity phases are utilised principally in open-tubular and support-coated open-tubular (SCOT) columns as when they are used in packed columns, saturated and monoenoic components of the same chain length are poorly separated.

It is occasionally necessary to subject fatty acids in the unesterified form to GLC analysis. In this instance, acidic liquid phases such as DEGS containing 2 per cent phosphoric acid,[345] or better carbowax 20 m–terephthalic acid[523] or the structurally related phase SP-1000 (Supelco Inc., U.S.A.), may be used.

The quality and to some extent the nature of the separations achieved is also influenced by the amount of liquid phase applied to the support. Low levels of polyesters (1–3 per cent by weight relative to that of the support) are occasionally suggested for use with highly inert supports as methyl esters of long-chain polyunsaturated fatty acids will elute from them at comparatively low temperatures. On the other hand, higher amounts of polyester (10–15 per cent by weight) offer greater protection to polyunsaturated esters so are generally preferred. The retention times of fatty acid esters relative to a chosen standard ester (usually 16:0 or 18:0) tend to decrease as the amount of liquid phase on the support is decreased so, for reproducible work, it is advisable to determine the optimum amount of liquid phase necessary for a given separation and to standardise one's chromatographic conditions accordingly. Unfortunately, as columns age with use, some changes in the retention characteristics of esters inevitably occur.

The nature of the support may also influence the quality of the separations (see Chapter 3). Acid-washed and silanised support materials, however, are almost completely inert so that separations are controlled largely by the liquid phase when these are used. A uniform fine grade (100–120 mesh is considered to be the optimum) is to be preferred for analytical columns. A good case could now be argued that all analyses of polyunsaturated fatty acids be performed on columns packed with 15 per cent EGSS-X or EGSS-Y on 100–120 mesh acid-washed or silanised supports except in special circumstances when they are not suitable.

There have been a number of reports of losses of esters of poly-unsaturated fatty acids on gas chromatographic columns. They can be attributed partly to the use of too active support materials and partly to transesterification of methyl esters with the polyester liquid phase. The latter effect is caused by residues of the catalyst required for the preparation of the polyester so that those components remaining longest on the columns will suffer the greatest losses. Fortunately, polyester liquid phases made without catalyst are now commercially available so that such losses need not be significant. If column packing materials are made with catalyst-free liquid phases coated on acid-washed silanised supports, quantitative recovery of most fatty acid esters is possible. Such materials are naturally more expensive, but it is a false economy to use inferior grades as columns will last for years if looked after properly. With open-tubular and SCOT columns, factors of this kind are less easily controlled and some losses on the walls of the columns are inevitable. Ackman[9] has discussed these problems.

(ii) *Provisional identification using standard mixtures*

Standard mixtures containing accurately known amounts of methyl esters of saturated, monoenoic and polyenoic fatty acids are commercially available from a number of reputable biochemical supply companies. These are invaluable for checking the quantification procedures used (see below) and also for the provisional identification of fatty acids by direct comparison of the retention times of their methyl esters with those of the unknown esters on the same columns under identical conditions. Comparisons should be made on at least two columns with different types of packing materials—one from group a above and the other from group b.

The lipids in animal tissues contain a much wider spectrum of fatty acids than is available commercially. It can therefore be helpful to obtain a secondary external reference standard consisting of a natural fatty acid mixture of known composition. This can be a common natural product that has been well characterised or a mixture of natural esters, the composition of which has been accurately established by the procedures described below. Ideally, it should be similar to the samples under investigation; for example, Ackman and Burgher[10] have used cod-liver oil in this way to identify the fatty acids of other marine animals and Holman *et al.*[216,217] have used the fatty acids of bovine and porcine testes for a similar purpose. Rat liver fatty acids also are frequently used in the same way. The author uses a simple mixture made up of the fatty acids of pig liver lipids and cod-liver oil together with a little linseed oil as this contains significant amounts of all the major fatty acid classes (saturated and mono-, di-, tri-, tetra-, penta- and hexaenoic components of both the (*n*-3) and (*n*-6) families and separable by silver nitrate chromatography), including most of the fatty acids likely to be encountered in animal tissues.

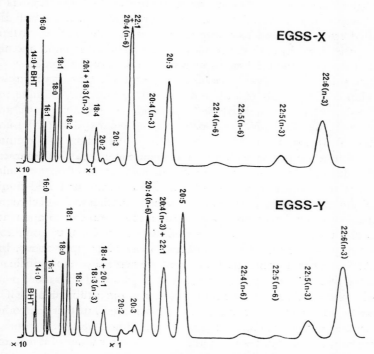

FIG. 5.1. Gas chromatographic analysis of a complex mixture of natural fatty acids (as the methyl esters) on columns packed with 15 per cent EGSS-X and EGSS-Y as stationary phases (see footnote to Table 5.1 for further chromatographic details).

Figure 5.1 shows the kind of separations of such a standard mixture, that can be obtained on 15 per cent EGSS-X and EGSS-Y packed columns (for full chromatographic details see the footnote to Table 5.1). Both columns give excellent separations of fatty acid esters of a given chain length that differ in degree of unsaturation. Separation of esters which differ only in the positions or configurations of the double bonds, where these are likely to be approximately central, is not easily achieved with monoenoic acids but is possible with esters of poly-unsaturated fatty acids. For example, on both columns, 18:3 (*n*-3) and 18:3 (*n*-6) are separated as are three isomers of 20:3, two isomers of 20:4 and two isomers of 22:5 as is apparent in Fig. 5.1. With the methyl esters of the more common families of polyunsaturated fatty acids, the shorter the distance between the last double bond and the end of the molecule the longer the retention time of the isomer.

The principal disadvantages of these columns is that there is some overlap of fatty acids of different chain lengths.[6] For example, on EGSS-X, 18:3(*n*-3) and 20:1(*n*-9) coincide as do 20:4 (*n*-6) and 22:1 (*n*-9). On EGSS-Y, these pairs can be separated but 20:0 and 18:3 (*n*-3), 18:4 (*n*-3) and 20:1 (*n*-9) or 22:1 (*n*-9) and 20:4 (*n*-3) cannot. By judicious use of both columns, most components can be separated and estimated, but it is necessary to use some form of open-tubular GLC or some other technique such as silver nitrate or liquid–liquid partition chromatography (see below) in combination with gas chroma-tography to completely separate, identify and estimate all components. It is also possible to eliminate the chain length problem by using low polarity polyester liquid phases such as those of group c above, but packed columns in which these are used are generally of low efficiency and the resolution of peaks is inferior to that obtained with more polar liquid phases. With group c liquid phases in open-tubular or SCOT columns, much greater resolution can be achieved but there are other difficulties with these columns (see Chapter 3 and below).

It should be noted, however, that the retention times of esters and the separations that can be achieved are all dependent on the precise column conditions used and may vary with the temperature or the age of a column and with the amount of stationary phase on the support.

(iii) *Relative retention times and equivalent chain lengths*

The absolute retention time of an ester on any gas chromatographic column has very little meaning as a measure of its elution character-istics because slight changes in the operating conditions or in the character of the packing material (in its origin or on ageing) can affect this parameter drastically. On the other hand, the retention time of a fatty acid ester relative to that of a chosen standard commonly occurring component (usually 16:0 or 18:0) has a greater absolute significance and is a quantity more suited to inter-laboratory comparisons, i.e.

$$\text{relative retention time } (r_{18:0}) = \frac{\text{retention time of ester}}{\text{retention time of 18:0}}$$

Theoretically, retention times should be measured from the time of injection of the sample into the gas chromatograph to the time when the peak is at its maximum, but as very large peaks may be skewed, it has been suggested that the distances should be measured to the point where the tangent drawn to the leading edge of the peak intercepts the base line.[3] This should rarely be necessary with modern packed columns but cannot be avoided with open-tubular columns. Also, as the retention times of esters are influenced to some extent by components eluting immediately adjacent to them, relative retention times should ideally be measured on pure compounds or on simpler fractions isolated by silver nitrate chromatography (see below) although this is not always possible. The relative retention times ($r_{18:0}$) of the component esters of the external reference standard, separated as illustrated in Fig. 5.1, on EGSS-X and EGSS-Y columns are listed in Table 5.1 (the precise operating conditions are listed in the legend to the table). Relative retention times also vary somewhat with the state of the column packing materials and with the operating conditions, but such variations are comparatively small and are in the same direction for all components.

Kovats retention indices are more generally accepted as a standard means of recording GLC retention data, but have been little used for the esters of fatty acids. Analogous parameters known as *equivalent chain lengths* (abbreviated to ECLs)[351] or carbon numbers[598] have considerable utility, however. ECL values are found by reference to the straight line obtained by plotting the logarithms of the retention times of a homologous series of straight chain saturated fatty acid methyl esters against the number of carbon atoms in the aliphatic chain of each acid. (Semilog. paper is particularly convenient for the

Lipid Analysis

TABLE 5.1. EQUIVALENT CHAIN LENGTHS AND RELATIVE RETENTION
TIMES OF SOME UNSATURATED ESTERS ON PACKED COLUMNS WITH
EGSS-X AND EGSS-Y AS STATIONARY PHASES

Methyl ester	ECLs		Relative retention times*	
	EGSS-X	EGSS-Y	EGSS-X	EGSS-Y
16:0	16·00	16·00	0·58	0·57
16:1 (n-9)	16·57	16·62	0·69	0·68
16:2 (n-6)	17·65	17·45	0·90	0·85
18:0	18·00	18·00	1·00	1·00
18:1 (n-9)	18·53	18·52	1·19	1·16
18:2 (n-6)	19·42	19·20	1·52	1·41
18:3 (n-6)	20·00	19·67	1·80	1·60
18:3 (n-3)	20·40	20·02	2·02	1·78
18:4 (n-3)	21·05	20·52	2·40	2·04
20:0	20·00	20·00	1·77	1·76
20:1 (n-9)	20·50	20·45	2·08	2·01
20:2 (n-6)	21·40	21·15	2·67	2·44
20:3 (n-9)	21·63	21·33	2·87	2·57
20:3 (n-6)	21·77	21·53	2·99	2·72
20:3 (n-3)	21·95	21·60	3·13	2·78
20:4 (n-6)	22·43	22·00	3·59	3·10
20:4 (n-3)	23·00	22·47	4·18	3·54
20:5 (n-3)	23·50	22·80	4·85	3·91
22:0	22·00	22·00	3·14	3·12
22:1 (n-9)	22·43	22·35	3·59	3·44
22:4 (n-6)	24·45	24·00	6·34	5·48
22:5 (n-6)	24·57	23·85	6·55	5·27
22:5 (n-3)	25·53	24·80	8·60	6·92
22:6 (n-3)	26·18	25·20	9·10	7·74
24:0	24·00	24·00	5·59	5·50

* Relative to 18:0.

Data obtained with 7 ft × ¼ inch o.d. glass columns packed with 15 per
cent (w/w) stationary phase on Chromosorb W (100–120 mesh, acid-
washed and silanised). Carrier gas—nitrogen at 50 ml/min. Column
temperatures, 194°C (EGSS-Y) and 178°C (EGSS-X).

purpose.) The retention times of the unknown acids are measured under
identical operating conditions and the ECL values are read directly
from the graph. The ECL values of the esters of the component acids
separated as illustrated in Fig. 5.1 were obtained in this way and are
also listed in Table 5.1. Such values have more obvious physical
meaning than relative retention times and are more easily remembered.

The increment in ECL value of a given ester over that of the saturated ester of the same chain length, sometimes known as the *fractional chain length* (or FCL) value, is dependent on the structure of the compound and is influenced by the number of double bonds in the aliphatic chain and the distance of the double bonds from the carboxyl and terminal ends of the molecule. The retention characteristics of the methyl esters of the complete series of monoenoic C_{18} acids [165] (i.e. Δ^2 to Δ^{17}), all the methylene interrupted *cis,cis*-dienoic acids[87] (i.e. $\Delta^{2,5}$ to $\Delta^{14,17}$) and for many acids with more than one methylene group between the double bonds[168] as well as of many natural polyunsaturated acids[208] have been obtained on a number of liquid phases so that the effect of double bond position on ECL values is now well documented. Jamieson[239] has collated these and many other results in a comprehensive review. A single double bond in the centre of a long aliphatic chain increases the ECL value over that of the corresponding saturated compound on a DEGS column by about 0·5, but as the double bond nears the carboxyl end of the molecule, this FCL value tends to decrease (except for the Δ^3 isomer) and as it nears the terminal end of the molecule, it tends to increase to a value of nearly 1·0 (the FCL value of the Δ^{16} isomer is even higher).[165] Proximity of a functional group of any kind to the terminal end of the molecule appears to have a greater effect on ECL values than when it is a similar distance from the carboxyl end. FCL values for individual double bonds are approximately additive; for example, if the data for the Δ^9 and [12] monoenes are used to predict the ECL value of the $\Delta^{9,12}$ diene, the calculated value is just a little lower than that obtained experimentally and the difference between the actual and calculated results is consistent over the entire methylene interrupted series.[87] Some interaction between the double bonds increasing the polarity of the molecule may be responsible for the minor discrepancy. When double bonds are separated by more than one methylene group, the difference between the calculated and experimental values is negligible.[168]

With any homologous series of fatty acids that contain a substituent in the alkyl chain, the distance of the substituent from either the proximal or the terminal end of the chain must vary and the logarithms of the retention times of the esters of such series plotted against the numbers of carbon atoms in the chain do not lie in a straight line unless the series are short. The deviations from a straight line are greatest for short-chain esters and with longer-chain compounds, the FCL values obtained for esters with similar groups of double bonds and terminal structures are approximately constant. Using the data in Table 5.1., on EGSS-X, 20:4 (*n*-6) has an ECL value of 22·43 (i.e. 20 +2·43) and that of 22:4 (*n*-6) is 24·45 (i.e. 22 +2·45); 20:4 (*n*-3) has an ECL value of 23·00 (i.e. 20 +3·00) so we can predict that 22:4 (*n*-3), which does not occur in the standard mixture, will have an ECL value of close to 25·00 (i.e. 22 +3·00). With a suitable secondary reference standard, the ECL values of the esters of most of the fatty acids likely to be encountered in animal tissues can be measured or predicted. Marine oils, which may contain more families of polyunsaturated fatty acids, present more complicated identification problems and there are difficulties in applying the factors to shorter-chain length esters. Ackman[3] has proposed the use of a series of systematic separation factors, which can be calculated from the relative retention times of esters, to assist in overcoming these problems.

Again it must be emphasised that ECL values may vary a little with column conditions (e.g. temperature and carrier gas flow rates), with the nature of the support, with the amount of liquid phase and with the age of columns. Such values obtained in allegedly similar circumstances in other laboratories may be taken as a guide but should be applied with caution.

The provisional identification of fatty acids can often be aided by ancillary techniques such as silver nitrate chromatography (see below).

(iv) *Open-tubular and SCOT columns*

The applications of open-tubular and SCOT columns to the analysis of fatty acids have recently been discussed in some detail by Ackman.[3] Positional and geometrical isomers of unsaturated fatty acid esters are much more easily separated by this means. For example, all the C_{18} *cis*-monoenoic isomers with the exception of the Δ^5 to Δ^9 compounds can be separated on open-tubular columns coated with ApL or NPGS.[165] Certain of the *cis*- and *trans*-isomers of these compounds can also be separated on the same columns but β-cyanoethylmethylsiloxane coated open-tubular columns apparently give better separations of geometrical isomers.[310,473] *Cis*-isomers are eluted before the corresponding *trans*-compounds on non-polar liquid phases and the reverse is true with polar liquid phases. Fortunately, ECL values obtained with packed columns are similar to those with open-tubular columns so that some comparisons between the two are possible. Overlap problems between compounds of different chain length tend to be much less serious with open-tubular columns,[6] but quantification can be a problem because of oxidation and retention of polyunsaturated fatty acids on the walls of the column.[13] Other problems associated with the use of these columns such as high cost and short working life have been discussed in Chapter 3. Ackman[8,9] has published some remarkable GLC traces of natural lipid samples illustrating how much more complex these are than analyses on conventional packed columns might indicate.

(v) *Quantitative estimation of fatty acid composition*

As discussed earlier (Chapter 3), with reliable modern gas chromatographs equipped with flame ionisation detectors, the areas under the peaks on the GLC traces are linearly proportional to the amount (by weight) of material eluting from the columns. Problems of measuring this area arise mainly when components are not completely separated and there is no way of overcoming this difficulty entirely. When overlapping peaks have distinct maxima, the height multiplied by retention time method of area measurement is probably the most accurate manual technique in isothermal analyses, but computer analysis of peak shapes[58] will improve the accuracy of the estimation. Where one component is visible only as a minor shoulder or broadening of a major peak, no manual or computer method is likely to give very precise results for the individual components although electronic integration will at least give an accurate measure of the total amount of material present. Wherever feasible, column conditions should be altered in an attempt to improve resolution or another liquid phase can be tried.

Other problems may arise through loss of components on the columns, but effects of this kind can be minimised with packed columns as discussed above by using catalyst-free liquid phases and highly inert supports, although it is necessary to check frequently whether losses are occurring by running standard mixtures of accurately known composition through the columns. It is essential that these standards should be similar to the samples to be analysed, for example saturated standards should not be used to calibrate columns for the analysis of polyunsaturated fatty acids, and that the calibration should be checked regularly. Methyl ester mixtures made to the specifications of the National Institutes of Health in the U.S.A. are available from several commercial sources. If necessary, calibration factors may have to be calculated for each component to correct the areas of appropriate peaks in the mixtures analysed. When flame ionisation detectors are used, small correction factors are sometimes applied to compensate for the fact that the carboxyl carbon atom in each ester is not combusted to a component to which the detector responds.[4,8,9,12] With open-tubular columns of all types, losses cannot be avoided and frequent calibration and calculation of correction factors is essential for quantitative work.[13] Good sample injection techniques can also improve the recoveries of fatty acids and the resolutions attainable. The syringe needle should be inserted rapidly but steadily into the column until it reaches its full length when the plunger is pressed in firmly; the needle is left in place for about 2 sec, then is removed rapidly.

Results can be expressed directly as weight percentages of the fatty acids present, but it is often necessary to calculate the molar amounts of each acid as, for example, in most lipid structural studies. This is performed simply by multiplying the area of each peak by an arithmetic factor obtained by dividing the weight of some standard ester (say 16:0) by the molecular weight of the component. It should be noted that if fatty acid compositions are calculated on a weight basis, it is not necessary to positively identify each compound, but this cannot be avoided if molar proportions are required.

2. *Unusual fatty acids*

Retention data, including ECL values, have been reported for the complete series of isomeric methyl-branched octadecanoates[2,5] and all are eluted before methyl nonadecanoate on polar and non-polar columns. The *iso-* and *anteiso*-isomers are those most often found in nature and they are easily separated from each other. The *iso*-compound is eluted first, and the C_{19} fatty ester with this structure has an ECL value of approximately 18·5 on 15 per cent EGSS-X and EGSS-Y while the related *anteiso*-compound has an ECL value of 18·7 on these columns (W. W. Christie, unpublished data). FCL values taken from the literature have been used to estimate ECL values for multibranched acids with some success.[7] Diastereoisomers of branched-chain acids have been separated by open-tubular gas chromatography.[7,11]

The ECL values of the complete series of methyl esters of C_{19} isomeric cyclopropane fatty acids have been recorded on polar and non-polar liquid phases[92] and they are all approximately one unit greater than those of the corresponding C_{18} monoenoic esters from which they are derived synthetically or biosynthetically. The 9,10- and 11,12-methyleneoctadecanoates, which are occasionally found together in bacterial lipids, can be separated on open-tubular columns.[92]

Tulloch[542] has reported ECL values for the complete series of methyl hydroxy-, acetoxy- and oxo-stearates on polar and non-polar liquid phases. As is usual with substituted compounds, components with central functional groups are not easily separated, but where the substituents are close to either end of the molecule, positional isomers can be resolved. A hydroxyl group increases the ECL value of a fatty acid considerably, especially on polar columns and on EGS; for example, methyl 12-hydroxystearate has an ECL value of 26·25 although its acetate derivative has an ECL value 1·5 units lower than this and the trimethylsilyl ether (TMS) derivative is eluted considerably more rapidly. Compounds with free hydroxyl groups present several problems to the gas chromatographer. Adsorption effects on the support or on the walls of the column and hydrogen-bonding effects may come into play so that unsymmetrical peaks are obtained and recoveries are incomplete and losses may also occur because of transesterification of the hydroxyl group with polyester liquid phases. Although such effects are less apparent with modern catalyst-free liquid phases and highly inert supports, it is still advisable to chromatograph the compounds in the form of a non-polar volatile derivative such as the acetate,[542] trimethylsilyl ether[593] or trifluoroacetate[588] derivatives as sharper peaks, better recoveries and better resolutions of positional

E

isomers are obtained while n-butylboronate derivatives[25,67] are invaluable for characterising 2- and 3-hydroxy fatty acids (see Chapter 4 for details of preparations). Polyhydroxy esters have also been subjected to GLC in the form of the trimethylsilyl ethers,[588] as trifluoroacetates[588] or as isopropylidene derivatives.[587] *Erythro-* and *threo-* forms of vicinal diols can be separated on packed columns when they are converted to either of the last two derivatives and as these compounds can be prepared quantitatively to a high degree of stereochemical purity from *cis-* or *trans*-olefins respectively, this provides a basis for gas chromatographic separation and estimation of stereoisomers of unsaturated acids on packed columns.[587] Epoxide derivatives can be used in a similar manner.[128,129] There appears to be a general rule that derivatives with the *trans-* or *threo*-configuration are eluted before those of the *cis-* or *erythro*-configuration, especially on non-polar stationary phases.

An isolated triple bond has a similar effect on the retention characteristics of a fatty acid as three methylene interrupted double bonds and methyl stearolate (methyl octadec-9-ynoate) is eluted with or slightly after methyl linolenate on DEGS[173] or PEGA[610] columns.

Methyl esters of cyclopropene fatty acids are less easily subjected to gas chromatography as they tend to decompose or rearrange on the columns to give spurious peaks, and it is more usual to modify them chemically by hydrogenation or by reaction with silver nitrate[245] or methanethiol[415] before analysis. If highly inert supports and silicone liquid phases are used, however, successful GLC of the native esters is possible;[422] the author has successfully chromatographed methyl sterculate on a well-conditioned EGSS-X column where it eluted just after methyl linoleate, although the same compound decomposed on a new EGSS-Y column. Other saturated and unsaturated cyclic esters are more stable to gas chromatography. Methyl 11-cyclohexylundecaanoate is eluted with 18:2 on a PEGA column and with 18:0 on ApL;[190] the cyclopentenoic acids of seed oils of *Hydnocarpus* species have also been subjected to GLC analysis.[611]

The presence of conjugated double bond systems in the alkyl chain increases the retention time of an ester considerably over that of a similar compound with methylene interrupted double bonds. In addition, the configuration of the double bonds has a much more pronounced effect and some separation of geometrical isomers of conjugated esters may be possible on packed columns.[105] Methyl 9-*cis*, 11-*trans*-octadecadienoate, for example, has ECL values of 20·48 on EGSS-X and 20·24 on EGSS-Y (W. W. Christie, unpublished data). Conjugated trienes reportedly undergo *cis–trans* isomerisation and double bond migration on polyester columns[346,358] and dehydration of compounds with allylic hydroxyl groups may occur leading to spurious peaks.[358] The latter problem is apparently not alleviated if the acetate or methoxy derivatives are prepared, although it might be instructive to repeat this work on more modern catalyst-free liquid phases on inert supports.

3. *Spurious peaks on recorder traces*

Spurious peaks appear on gas chromatographic traces when fatty acids, such as those with cyclopropene rings or with hydroxyl groups adjacent to double bonds, rearrange or decompose as detailed above. Other troublesome artefacts may arise during the methylation process as methoxy-compounds are sometimes formed from unsaturated fatty acids when boron trifluoride–methanol[143,312] or sulphuric acid–methanol[176,191] are used as esterifying reagents and polymeric by-products may be formed if diazomethane[365,461] or dimethoxypropane and methanolic hydrogen chloride[365] are used. Difficulties of this kind need not occur if the procedures described in Chapter 4 are selected with care. If cholesteryl esters are transesterified directly and the free cholesterol is not removed prior to GLC analysis, it may dehydrate to form cholestadiene on columns made up of low percentages of EGSS-X or EGSS-Y and obscure the C_{22} polyenoic esters.[286] With higher amounts of liquid phase this does not occur and cholesterol is eluted so late that it does not significantly affect the base line and can be ignored. The methyl ether derivative of cholesterol is sometimes formed with acidic reagents.[366] Other potential contaminants can appear in any of the reagents used in lipid analysis as has been discussed in Chapter 3. Vigilance must be exercised continuously to detect and eliminate them.

C. Isolation of Individual Fatty Acids for Structural Analysis

It is often necessary to isolate fatty acids in a pure state before they can be unequivocally identified. Adsorption chromatography is invaluable for the isolation of fatty acids with polar functional groups, but silver nitrate chromatography, partition chromatography or combinations of both may be required to separate individual unsaturated fatty acids. The use of urea to obtain concentrates of polyunsaturated or branched-chain acids is discussed in section I.2 below.

1. *Adsorption chromatography*

In general, fatty acids that differ only in chain length or degree of unsaturation cannot be separated by adsorption chromatography although short-chain and polyunsaturated fatty acids migrate more slowly on silica gel and other adsorbents than C_{16} to C_{20} saturated or monoenoic components and fractions enriched in these compounds can sometimes be obtained. If polar functional groups occur in the alkyl chain, however, some useful separations are possible. It is again more usual to separate fatty acids as the methyl ester derivatives, but unesterified fatty acids can also be chromatographed if one per cent of acetic or formic acids are incorporated into the solvent systems when silicic acid is the adsorbent. Sufficient material for many structural analyses can be obtained by preparative TLC and the elution characteristics of polar fatty acid esters on thin layers of silica gel have been thoroughly examined.

In particular, Morris *et al.*[363] have studied the chromatographic behaviour of the complete series of methyl hydroxy-, acetoxy- and keto-stearates. When each series is arranged in order as a line of spots on a single TLC plate which is then developed in various hexane–diethyl ether solvent mixtures, the compounds emerge as a sinusoidal curve with a minimum at the 5-substituted isomer and a maximum at the 13- or 14-substituted isomer. Isomers with functional groups in positions 2 to 7 can often be separated from each other and from the remaining compounds. This sinusoidal effect is exhibited by other isomeric series of polar aliphatic compounds. Methyl 6-hydroxy-, 14-

hydroxy- and 2-hydroxy-stearates have been separated by TLC on silica gel layers[558] with hexane–diethyl ether (85:15, v/v) as solvent system. Figure 5.2 illustrates the elution characteristics of some hydroxy esters with this solvent system. Positional isomers of dihydroxy esters, i.e. methyl 6,7-dihydroxy- and 9,10-dihydroxy-stearate have also been separated by TLC with diethyl ether–hexane (4:1, v/v) as developing solvent.[361] Up to 20 mg of esters can be separated on a 20×20 cm plate coated with a layer of silica gel 0·5 mm thick. Bands are detected by means of the 2',7'-dichlorofluorescein spray and recovered by eluting the adsorbent in a column with diethyl ether or chloroform–methanol mixtures (about 9:1, v/v).

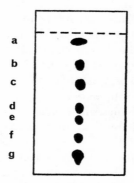

FIG. 5.2. Thin-layer chromatography of methyl esters of oxygenated fatty acids on silica gel G. Solvent system, hexane-diethyl ether (85:15, v/v). a, methyl palmitate; b, methyl 9,10-epoxystearate; c, methyl 12-keto stearate; d, methyl 2-hydroxy stearate; e, methyl 14-hydroxy stearate; f, methyl 6-hydroxy stearate; g, methyl 9,10-dihydroxy stearate.

Column chromatography on silicic acid has also been used to separate polar fatty acid esters (Radin[414] has reviewed some applications) and methyl 9-hydroxy- and 13-hydroxy-stearates[122] and the acetate derivatives of methyl 9,10-dihydroxy-myristate and 9,10-dihydroxy-palmitate[94] have been resolved by this technique.

Threo- and *erythro*-isomers of vicinal dihydroxy-esters can be separated on thin layers of silica gel impregnated with boric acid[355] (see Chapter

3) with hexane–diethyl ether (60:40, v/v) as developing solvent (the *threo*-isomer migrates farthest), although the compounds cannot be separated on silica gel alone. Sodium arsenite impregnated layers give even more remarkable separations of diastereoisomers.

Where the fatty acids contain double bonds in addition to more polar groups, silver nitrate chromatography may be of additional assistance in achieving separations (see below).

2. *Silver nitrate chromatography*

Since its introduction by Morris in 1962, TLC on silica gel impregnated with silver nitrate has been of enormous value to the lipid analyst. Fatty acids can be separated according to both the number and the configuration of their double bonds and often with care, according to the position of the double bonds in the alkyl chain. Hexane–diethyl ether solvent systems are useful for separating general classes of methyl esters as illustrated in Fig. 5.3, although it is not possible to resolve components with three or more double bonds together with those with zero to two double bonds on a single plate. Indeed, complete separation of components with more than four double bonds is never easy. Hexane–diethyl ether (90:10, v/v) will separate components with up to two double bonds and the same solvents in the ratio 40:60 will separate polyunsaturated esters. Bands are visualised under UV light after spraying with 2′,7′-dichlorofluorescein solution and components with zero to two double bonds are eluted from the adsorbent with diethyl ether or chloroform; chloroform–methanol (9:1 v/v) may be necessary for complete recovery of polyunsaturated compounds. Methods of eliminating silver ions from the extracts are discussed in Chapter 3. Up to 10 mg of esters can be separated on a 20 × 20 cm plate coated with a layer 0·5 mm thick of silica gel containing 10 per cent (w/w) silver nitrate. The GLC recorder tracings on the EGSS-X column of fractions obtained from the GLC reference mixture (Fig. 5.1) on these plates are also illustrated in Fig. 5.3. It is easy to see that considerable simplification of the sample has been achieved and that the separations are consistent with the tentative identifications made for each component.

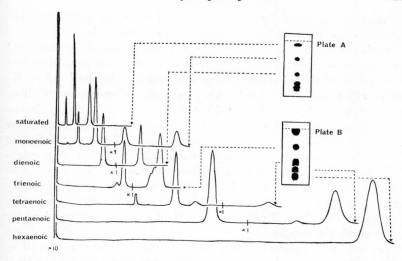

FIG. 5.3. GLC recorder traces (the EGSS-X column of Table 5.1 and Fig. 5.1) of esters separated into saturated, mono-, di-, tri-, tetra-, penta- and hexaenoic fractions by TLC on silica gel G impregnated with 10 per cent (w/w) silver nitrate. A, solvent system, hexane–diethyl ether (90:10, v/v); B, solvent system, hexane–diethyl ether (40:60, v/v).

The elution characteristics of a wide variety of unsaturated esters have been studied. For example, the complete series of methyl *cis-* and *trans*-octadecenoates[165] and methylene interrupted *cis,cis*-octadecadienoates[86] have been subjected to silver nitrate TLC. When run in order on a single TLC plate, each series migrates in the form of a sinusoidal curve similar to that observed with polar esters on ordinary silica gel, with a minimum at the Δ^6 isomer ($\Delta^{6,9}$ in the case of the dienes) and a maximum at the Δ^{13} isomer. *Trans*-isomers migrate consistently ahead of the corresponding *cis*-isomers with the exception of the *cis*-2 component which not only migrates ahead of its *trans*-analogue but also ahead of methyl stearate. Natural monoenoic fatty acids containing *trans*-double bonds can be estimated by separating them from the *cis*-compounds by means of silver nitrate TLC with hexane–diethyl ether (9:1, v/v) as developing solvent and eluting them together with any saturated components from the adsorbent. If the samples are analysed by GLC before and after the separation, the amount of the *trans*-acids in the mixture can be determined.[98]

With care it is possible to separate positional isomers of unsaturated fatty acids by silver nitrate TLC. The most consistent and successful separations of this kind have been achieved by Morris *et al.*[362] who utilise silica gel impregnated with very high concentrations of silver nitrate (up to 30 per cent) and develop the plates several times in the same direction if necessary with toluene as developing solvent at low temperatures ($-5°$ to $-25°C$). With this system, the methyl 6-, 9- and 11-*cis*-octadecenoates are well separated. Such separations are particularly useful in that the first two of these fatty acids occur together in many seed oils of the family Umbelliferae and cannot be resolved by any other chromatographic procedure. Layers containing such high proportions of silver nitrate are very friable but are not required for more routine separations.

An acetylenic group is less polar than a *cis*-double bond on silver nitrate TLC so methyl stearolate migrates just ahead of methyl oleate[359] and methyl crepenynate ahead of methyl linoleate.[167] The allenic ester, methyl labellenate, also migrates ahead of methyl oleate.[168] Conjugated double bonds are less strongly retained than similar compounds with isolated double bonds; for example, methyl 9-*cis*, 11-*trans*-octadecadienoate migrates with methyl oleate when hexane–diethyl ether (9:1, v/v) is the developing solvent and just ahead of it when toluene is the developing solvent (W. W. Christie, unpublished work). It is worth noting that compounds which are not resolved with hexane–diethyl ether solvent systems are frequently separable with aromatic solvents such as toluene and vice versa. Cyclopropene fatty acids, however, are destroyed by silver nitrate chromatography and cannot be purified by this means although the reaction has been used in a method of estimating them.[245] Cyclopropane and saturated branched-chain esters elute with normal saturated straight-chain compounds.

Larger amounts of esters can be separated by column chromatography on silver nitrate impregnated adsorbents of which the most useful is acid-washed Florisil containing 16 per cent (w/w) silver nitrate (see Chapter 3). Up to a gram of esters can be separated on 40 g of adsorbent; saturated esters elute with hexane–diethyl ether (199:1, v/v), *cis*-monoenes with the same solvents in the ratio 99:1 and dienes with the solvents in the ratio 94:6.[583]

Before silver nitrate chromatography was developed, some similar separations of unsaturated compounds were achieved by thin-layer chromatography of mercuric acetate derivatives. The procedure is rather tedious as the derivatives must be prepared prior to the analysis (see Chapter 4), then decomposed when the fractionation is complete before components can be analysed further by other procedures. Also, the resolutions that can be obtained are not as good as those achieved with silver nitrate chromatography (the topic has been reviewed by Mangold[335]).

3. *Preparative gas chromatography*

Gas chromatography was developed initially as an analytical tool, but many commercial instruments can be adapted to permit analytical columns to be used preparatively. Specially designed equipment is also available that can be used to purify gram quantities of methyl esters, but sufficient material can usually be obtained from the simpler analytical instruments for fatty acid structures to be determined by microchemical methods.

If an analytical column is to be used preparatively, the gas chromatograph must have certain essential features. These include an efficient flash heater to ensure that the larger amounts of material to be separated are instantly volatilised and enter the column as a narrow band, a stream splitter at the exit end of the column (if the instrument has a flame ionisation detector) so that only 1 per cent or so of the column effluent is destroyed by the detection system although a continuous recorder trace of the components is obtained as they elute, and finally, the outlet from the stream splitter to a collection system should be heated so that components do not condense prematurely and contaminate the outlet. Conventional $\frac{1}{4}$ inch o.d. analytical columns may be used to separate 1–2 mg quantities of esters preparatively, but a considerable loss of resolution will occur and better results can be obtained if very high amounts of liquid phase (30 per cent or so) are used on a slightly coarser support (60–80 mesh) than is usual in analytical columns. If the oven dimensions are suitable, $\frac{3}{8}$ inch o.d. columns, which have a much greater capacity than those of $\frac{1}{4}$ inch, may be used to purify larger amounts of esters.

The precise amounts that can be conveniently separated preparatively depend partly on the nature of the column and partly on the nature of the sample. If the latter is a complex natural mixture containing a number of saturated and unsaturated components, it will be necessary to use a polyester liquid phase and inject only 2–3 mg at a time to ensure adequate resolution. The process may have to be repeated several times with pooling of corresponding fractions to obtain sufficient material for structural analyses. Several difficulties may become apparent. With larger amounts of polyester liquid phase in wider columns, it is generally necessary to heat the columns to a temperature close to the natural thermal limit of the polyester in order to elute components in a reasonable time and as a result fractions may be contaminated by liquid phase bleeding from the columns. Such impurities can be removed later by TLC on silica gel G (hexane–diethyl ether in the ratio 95:5 by vol. is a suitable developing solvent), but this introduces an extra step that may compound any other losses.

A more useful procedure is to simplify the mixture by silver nitrate or liquid–liquid partition chromatography and to separately subject the fractions obtained to preparative gas chromatography. By this means, the resolution problem is considerably lessened, low-bleed temperature-stable silicone liquid phases can be used in the column packings and much larger amounts of material (5 mg or so of each ester on $\frac{3}{8}$ inch columns) can be injected. Indeed, as a precise quantitative record may not be required, it is often possible to separate 20 mg or more of each component for, although highly unsymmetrical peaks will be obtained on the recorder tracing, the resolving powers of the columns will probably be sufficient to separate compounds differing in chain length by two carbon atoms cleanly. Polyester liquid phases must be used, however, if separation of positional isomers of poly-unsaturated esters is required.

The collection of methyl esters issuing from the columns also presents certain problems as on rapid cooling they tend to form aerosols that are not easily condensed and losses can be high. Most manufacturers of GLC equipment supply specially designed traps which lead the effluent through a tortuous path so that the vapour has a greater opportunity to come in contact with the sides of the trap and to stick, but very long straight tubes are often as effective as these. A highly efficient yet simple trapping device has been described by Schlenk and Sand[463] which combines alternate heating and cooling of a long tube, but electrostatic precipitation[54] may be even more effective. Better recoveries can also be obtained if the carrier gas flow is interrupted by passing it through a short distance of column packing material on which the sample will condense and from which it can later be washed with a suitable solvent.[260] Millipore filters can also be used to collect small samples efficiently.[178] If a small amount of hexane is injected into the column immediately before a peak is collected, any material that has condensed prematurely from earlier fractions will be removed and will not cause any contamination of the next peak.[416]

There remains a slight possibility that methyl esters of polyunsaturated esters may be altered during chromatography at high temperatures so will not be suitable for further structural analyses. The evidence is somewhat contradictory in that there are reports that no changes occur and others, equally authoritative, that esters such as 22:5 and 22:6 undergo isomerisation similar to that which occurs during distillation of these compounds. It is possible that the design of the gas chromatographs used for the purpose, particularly at the outlet end, may have some significance with regard to the problem, but, whatever the reason, esters with fewer than three methylene-interrupted double bonds are unlikely to be affected. Preparative gas chromatography has been reviewed elsewhere.[199,200]

4. *Liquid–liquid partition chromatography*

Before gas chromatography was developed, liquid–liquid partition chromatography was the most useful technique for separating individual (or critical pairs of) fatty acids from natural mixtures. Lately the procedure has gone out of fashion despite the wide range of sophisticated equipment, including liquid chromatography detectors, controlled porosity supports, liquid phases chemically bound to supports to eliminate problems of bleeding and high pressure solvent delivery systems, that have become available commercially in the last few years. It will be surprising if the technique does not return to favour soon, especially for the separation of isotopically labelled fatty acids.

The reverse phase column chromatographic system developed by Privett and Nickell[410] in which heptane is supported as the stationary phase on silanised celite and acetonitrile–methanol (85:15, v/v) is the mobile phase is the most stable, reproducible and, therefore, widely used method and can be particularly recommended. Both phases are volatile and can be used with numerous liquid detection systems while the column is readily regenerated and can be used repeatedly. The column is packed with 100-mesh celite made hydrophobic by treatment with dimethyldichlorosilane, then heptane saturated with acetonitrile–methanol (85:15, v/v) is passed through the column and when it reaches the bottom, the acetonitrile–methanol phase saturated with heptane is allowed to flow through the column. When this emerges at the bottom, the sample is placed on top of the column and allowed to percolate in a few cm before the solvent flow is restarted. Methyl laurate emerges in 8 column volumes, methyl myristate in 11 and methyl palmitate in 16. Up to 2 g of methyl esters can be applied to a $120 \times 2 \cdot 5$ cm column.

Compounds as they elute can be detected by removing the solvents and weighing or by oxidising them with chromic acid (see Chapter 3) or continuously by means of a differential refractometer or by a moving wire detection system. Indeed, with this last detector, with very narrow bore columns and high pressure solvent delivery pumps, it should theoretically be possible to use the system as an analytical or small-scale preparative tool with a high degree of resolution (possibly even of critical pairs).

Other efficient solvent systems, including one with a silicone stationary phase and acetone–water as mobile phase[462] and others with powdered rubber[53,203] as the support medium and stationary phase, have been supplanted by Privett's procedure, which also has considerable potential as a large-scale preparative tool.

Thin-layer liquid–liquid partition systems, operating on the reverse phase principle, are useful for analytical or small-scale preparative purposes and free fatty acids or their methyl esters are readily separated into critical pairs. Silicone oil (Dow Corning 200 fluid) has been the most popular stationary phase, though undecane and tetradecane are also used, with acetonitrile–acetic acid–water mixtures as the mobile phase (for example methyl esters can be separated with these solvents in the proportions, 70:10:25 by vol.).[334] Further details of alternative solvent systems are available elsewhere.[333,335] Normally, samples are applied as discrete spots or as streaks in a volatile solvent,

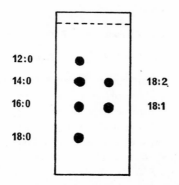

FIG. 5.4. Reverse phase separation of methyl esters on thin layers of siliconised silica gel G with acetonitrile–methanol–water (6:3:1, by vol.) as mobile phase.[389]

up to 10 mg of esters preparatively, to a 20×20 cm plate and the separation is complete in 2–3 hr. The principal disadvantage of the technique is that it is somewhat messy and that components are difficult to detect. The procedure of Ord and Bamford[389] overcomes the first of these drawbacks; an ordinary silica gel G TLC plate is silanised by standing it in a tank of dimethyldichlorosilane vapour for several hours. The methyl groups, which are then chemically bonded to the silica gel, serve as the non-polar stationary phase on which the fatty acids or esters are separated with acetonitrile–methanol–water (6:3:1 by vol.) as mobile phase. Figure 5.4 illustrates the kind of separation possible. For analytical purposes, 20 μl spots are applied to the plate and detected with a phosphomolybdic acid spray. Preparatively, 1–2 mg of esters can be separated on a 20×20 cm plate, but the edges of the plate must be exposed to iodine to detect the bands and as they may not always be straight and parallel, some loss may occur.

Liquid–liquid chromatography does not in general give pure single components from natural mixtures but rather critical pairs (see Chapter 3). As such components differ both in degree of unsaturation and chain length, the fractions can then be subjected to silver nitrate chromatography or to preparative gas chromatography so that specific pure compounds are ultimately obtained by the combined techniques.

D. Spectroscopy of Fatty Acids

1. *Infrared absorption spectroscopy*

Fatty acids may be subjected to IR spectroscopy in the unesterified state, when bound to glycerol or as the methyl esters, although an esterified form is to be preferred as a band due to the free carboxyl group between 10 and 11 μm might otherwise obscure a number of other important features in the spectra. Most information on the chemical nature of fatty acid derivatives can be obtained when they are in solution (see Chapter 3) and Fig. 5.5 illustrates the IR spectrum of soyabean oil in carbon tetrachloride solution. The sharp band at 5·75 μm is due to the esterified carbonyl function which is also responsible for a band at 8.9μm. With free fatty acids, the first of these bands is displaced to 5·9 μm and there are also broad bands at 3·5 μm and 10·7 μm. *cis*-Double bonds give rise to small bands at 3·3 μm and 6·1 μm and, although they are useful as diagnostic aids, they are not sufficiently distinct to be of use in quantitative estimations. Most of the remaining bands are absorption frequencies of the hydrocarbon chain.

A number of other functional groups give rise to characteristic bands that can be used to identify or to estimate the amount of a given acid. The most important of these is a sharp peak given by *trans*-double bonds at 10·3 μm and IR spectroscopy has long been the principal method for estimating compounds with this structural feature in the fatty acid chain. In the recommended procedure,[534] a standard is run at the same time as the unknown and the amount of *trans*-isomer in the latter is calculated from the ratio of the absorptivities at 10·3 μm. In

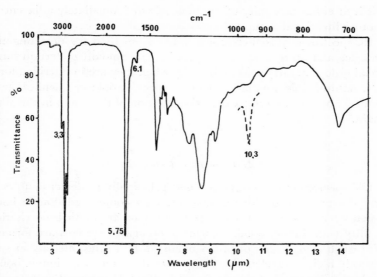

Fig. 5.5. Infrared spectrum of soyabean oil in carbon tetrachloride solution. The insert at 10·3 μm illustrates the absorption band due to a *trans*-double bond.

an alternative rapid procedure, the ratio of the absorbances at 8·6 μm (due to the carbonyl function) and at 10·3 μm in the unknown is measured and the amount of *trans*-isomer present is obtained by reference to a standard curve.[18] For accuracy with this last procedure, however, the standard curve must be prepared with material very similar to that to be analysed.[233] If a *trans*-double bond is part of a conjugated system, the band maximum may be shifted to about 10·1 μm.

Free hydroxyl groups give rise to a sharp band at 10·9 μm, epoxyl groups produce a double band with maxima at 11·8 and 12·1 μm, allenes give a small band at 5·1 μm, a cyclopropene ring produces a small band at 9·9 μm and a cyclopropane ring produces two small bands at 3·25 and 9·8 μm. Some of the absorption frequencies caused by the more unusual functional groups have been discussed by Wolff and Miwa;[585] more detailed general information can be found in a number of reviews.[82,83,140,387] Characteristic frequencies are not altered markedly by the position of a group in the aliphatic chain unless this is at either extremity of the molecule or immediately adjacent to another functional group.

IR spectroscopy is therefore a valuable method for detecting the presence and on occasion for estimating the amount of certain functional groups in fatty acids, especially when they are in esterified form in a natural lipid mixture but other spectroscopic or chemical procedures must be used to fix the exact position of the group in the alkyl chain.

2. *Ultraviolet spectroscopy*

UV spectroscopy is nowadays used principally to detect or to confirm the presence of fatty acids containing conjugated double bond systems in natural oils, or to detect chemical or enzymatic isomerisation of fatty acid double bonds in which conjugated systems are formed. With such acids, series of broad bands of increasing intensity at successively higher wavelengths, the greater the number of double bonds in the conjugated system, are found in the UV region. For example, with conjugated dienes, λ_{max} is at 232 nm ($\varepsilon = 33,000$), with conjugated trienes (e.g. α-eleostearic acid), λ_{max} is at 270 nm ($\varepsilon = 49,000$) and with conjugated tetraenes (e.g. α-parinaric acid), λ_{max} is at 302 nm ($\varepsilon = 77,000$). Different geometrical isomers have slightly different spectra; the greater the number of *trans*-double bonds, the higher the extinction coefficient but the shorter the wavelength of the band maxima. Conjugated triple bonds also affect the spectra. The absorbance at 234 nm is used to detect the formation of the *cis-trans*-conjugated double bond system produced by the action of lipoxygenase.[212] More detailed reviews of the applications of UV spectroscopy exist.[83,397]

3. *Nuclear magnetic resonance spectroscopy*

In recent years, NMR spectroscopy has been increasingly applied to the identification of lipid structures and in particular to the identification and often to the location of the double bond systems in fatty acid chains. Methyl esters are generally prepared for analysis and Fig. 5.6

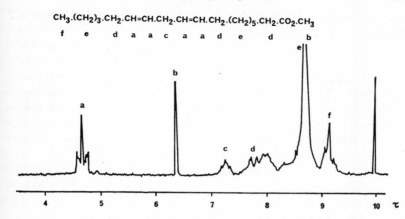

$CH_3.(CH_2)_3.CH_2.CH=CH.CH_2.CH=CH.CH_2.(CH_2)_5.CH_2.CO_2.CH_3$

FIG. 5.6. NMR spectrum of methyl linoleate in carbon tetrachloride solution (15 per cent, w/v) at 60 MHz (tetramethylsilane as internal standard).

illustrates the NMR spectrum of methyl linoleate. There are six main features: a multiplet at 4·72 τ for the olefinic protons, a sharp singlet at 6·40 τ for the protons on the methoxyl group, a triplet at 7·25 τ for the methylene group between the double bonds, a complex multiplet at 7·88 τ for the protons α to the double bonds and carboxyl group, a broad band at 8·67 τ for the chain methylene protons and finally a partially resolved triplet at 9·10 τ for the terminal methyl protons.[563] A wide range of unsaturated fatty acids have been subjected to NMR spectroscopy on 60 MHz instruments. They include the complete series of methyl octadecenoates,[164] all the methylene interrupted *cis*, *cis*-octadecadienoates,[95] several *cis*, *cis*-octadecadienoates with more than one methylene group between the double bonds[169] and large numbers of natural polyunsaturated fatty acids.[163,223] From studies with such compounds, it is now known how variations in the positions of double bonds affect the NMR spectrum of a long-chain fatty acid. For example, the Δ^2 to Δ^5 and Δ^{14} to Δ^{17} monoenoic isomers can be distinguished by this technique largely because of small changes in the signal associated with the olefinic protons.[164] With a more powerful instrument (220 MHz) only the Δ^{10} and Δ^{11}-octadecenoates cannot be distinguished.[142] The configuration of the double bond has little effect on the spectrum and stereoisomers are not easily identified by NMR spectroscopy. The terminal methyl group of polyunsaturated fatty acids of the (*n*-3) series produce a well-separated triplet at a slightly lower field (9·02 τ) than usual and this feature has been utilised in the estimation of such compounds in the natural mixtures.[155] The intensity of the band due to the methylene group between the double bonds can be used to determine the degree of unsaturation of a given fatty acid or a natural oil. NMR spectra of some natural conjugated trienoic acids (*α*- and *β*-eleostearic acid) have been determined; the olefinic protons give rise to a complex multiplet at approximately 4·0 τ.[223]

Free hydroxyl groups in fatty acid chains give rise to two separate signals; that due to the –OH proton is indistinct and may vary because of hydrogen bonding effects, but that due to the –CHO– proton at 6·4 τ is quite characteristic. All the isomeric hydroxy stearates have been examined by NMR spectroscopy and all can be distinguished from each other by this technique when quinoline is used as solvent.[543] When the hydroxy esters are acetylated, the acetoxy protons give rise to a sharp signal at 7·9 τ. Keto groups influence the *α*-methylene protons which produce a signal similar to that for protons adjacent to a carboxyl group. Many other functional groups give rise to distinctive signals; epoxide ring protons at 7·2 τ,[225] cyclopropene ring protons at 9·2 τ,[225] cyclopropane ring protons at 9·4 and 10·3 τ[311] and olefinic protons in a cyclopentene ring at 4·3 τ.[223] Although an acetylenic bond does not possess protons (unless it is terminal), the signals of protons on the *α*-methylene group can be used to identify it and when it is near either end of the molecule, to fix its position.[164] Methyl branches on aliphatic chains, however, do not give signals that are helpful in locating their positions unless the branch is immediately adjacent to either end of the molecule; *iso*-compounds can, therefore, be recognised, but *anteiso*-compounds cannot be distinguished from fatty acids having the methyl group in a central position.

Methyl esters are generally dissolved in carbon tetrachloride for analysis by NMR spectroscopy, but chloroform, although it produces a single peak at 2·75 τ, or deuterated solvents may also be used. The application of NMR spectroscopy to fatty acid analysis has been reviewed comprehensively elsewhere.[163,223,224] As more powerful instruments become available, the technique will undoubtedly become an even more valuable tool but it is used to greatest advantage on individual pure compounds, isolated by the procedures described above, rather than on natural mixtures.

4. *Mass spectrometry*

A wide variety of natural and synthetic fatty acids, usually in the form of the methyl ester derivatives, have been studied by mass spectrometry. Long chain saturated esters are easily identified[450] and are characterised by a prominent molecular ion (M^+) and other significant ions at $m/e = $ M-31 (loss of methanol) and M-43 (loss of C_2, C_3 and C_4 as a result of a complex rearrangement) together with a series of ions of general formula $-CH_3O_2C.(CH_2)_n^+$. The base peak at $m/e = 74$ is known as the 'McLafferty rearrangement ion" and is formed in a rearrangement reaction after cleavage of the parent molecule beta to the carboxyl group.[326]

Mass spectra of unsaturated fatty acids differ considerably from those of the corresponding saturated compounds but are less useful for characterisation purposes. The molecular ion is often prominent and there are generally major fragments at $m/e = $ M-31 or M-32 (loss of methanol) and, with monoenes and dienes, at $m/e = $ M-74 (loss of the McLafferty rearrangement ion).[181] There are no ions that can be used to locate the position of double bonds, which probably migrate when the molecule is ionised unfortunately, although there are indications that some positional isomers of polyunsaturated fatty acids may have distinctive but not easily interpreted spectra.[217] Double bond positions can be determined, however, if the unsaturated esters are first converted to a suitable derivative, preferably one which is sufficiently volatile to be subjected to gas chromatography. A wide variety of such derivatives have been tried—usually the double bond is oxidised to an epoxide or to a vicinal diol which is then reacted further to form more stable or more volatile compounds. The most useful derivatives have proved to be vicinal diols in the form of the isopropylidene[322] or trimethylsilyl ether[28,72] derivatives (see Chapter 4 for methods of preparation). The former permits the configuration of the double bond to be established because of the different GLC retention characteristics

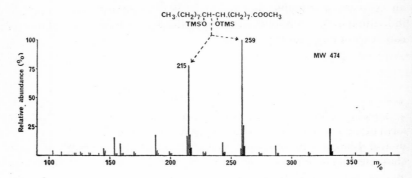

Fig. 5.7. Mass spectrum of the trimethylsilyl ether derivative of the vicinal diol
prepared from methyl oleate.

of the stereoisomers; the latter gives more intense fragments that can
be related to the position of the original double bond. Figure 5.7
illustrates the mass spectrum of the trimethylsilyl ether of the vicinal
diol prepared from methyl oleate. Ions in the high mass range are of
very low intensity and the molecular ion does not appear, but two ions
at $m/e = 215$ and $m/e = 259$, which represent cleavage between the
two carbon atoms that originally constituted the double bond, are
particularly prominent. Indeed, these ions are sufficiently intense to
be of use in estimating mixtures of positional isomers of unsaturated
fatty acids. Trimethylsilyl ether derivatives prepared from polyunsaturated acids have also been characterised by mass spectrometry
although the methyl ethers of polyols offer some advantages as suitable
derivatives in this instance.[380]

Similarly, acetylenic bonds in fatty acids cannot be located by mass
spectrometry unless they are converted to a suitable derivative. In this
instance, the elements of water are added to the bond by acidic hydrolysis of the mercuric acetate derivative (see Chapter 4) after which the
structure of the resulting ketone is determined by mass spectrometry.[71]

Most oxygenated fatty acids can be recognised and identified by mass
spectrometry as there are usually prominent ions formed by cleavage
beta to the oxygen function.[453] Methyl-branched esters have spectra
similar to the corresponding straight-chain esters with the same numbers

of carbon atoms but changes in the intensities of some ions can be used to indicate the presence and fix the position of the branch,[451,452] although the spectra are more readily interpreted if the esters are first reduced to the alcohols with lithium aluminium hydride. Cyclopropane esters are not readily distinguished from mono-unsaturated esters with similar numbers of carbon atoms by mass spectrometry, apparently because on ionisation the cyclopropane ring opens to form such a monoenoic compound. It is therefore necessary to disrupt the ring by vigorous hydrogenation so that esters with methyl branches and characteristic mass spectra are formed.[321] Alternatively methoxy derivatives can be prepared by reaction of the cyclopropane ester with boron trifluoride-methanol for mass spectral identification.[347,348] Cyclopropene esters are apparently too labile to be subjected directly to mass spectrometry, but the position of the ring can be located by this technique if the compound is oxidised to a β-diketone[222] or reacted with methanethiol to form a product with the sulphur atom attached to either of the ring carbons.[416] Other larger ring cyclic fatty acids such as those with cyclopentene,[100] cyclohexane[472] or furanoid[360] groups give quite distinct spectra from which the structures are readily deduced.

Mass spectrometry of fatty acids and their derivatives have been comprehensively reviewed by McCloskey.[319,320] Although mass spectrometry cannot always be used to identify a given fatty acid, it normally gives the molecular weight and some other basic information about the nature of the compound.

E. Hydrogenation

One of the first steps in the determination of the structure of an unsaturated fatty acid is to establish its chain length and this can be ascertained simply by means of catalytic hydrogenation to the saturated compound which is then identified positively by gas chromatography. Many of the published hydrogenation procedures are needlessly complex and the following method is adequate for most purposes.

"The unsaturated ester (1–2 mg) is dissolved in methanol (1 ml) in a test-tube and Adams' catalyst (platinum oxide; 1 mg) is added. The tube is connected via a two-way tap to a reservoir of hydrogen at atmospheric pressure and to a water-pump. The tube is alternatively evacuated and flushed with hydrogen several times to remove any air, then it is shaken vigorously while an atmosphere of hydrogen at a slight positive pressure is maintained for 2 hr. At the end of this time, the hydrogen supply is disconnected, the tube is flushed with nitrogen and the solution is filtered to remove the catalyst. The solvent is removed under reduced pressure and the required saturated ester is taken up in hexane or diethyl ether for GLC analysis."

Hexane may be used as solvent for the hydrogenation reaction if the fatty acid is still esterified to glycerol as in a triglyceride or diglyceride acetate, but the hydrogenated compounds must later be recovered from the catalyst with a more polar solvent such as chloroform.

In the above method, the amount of hydrogen taken up is not measured but this can be done in order that the number of double bonds in the molecule be accurately determined if an apparatus constructed for the purpose is used. Alternatively, the procedure of Brown et al.,[68] in which the platinum catalyst is generated *in situ* by treatment of platinum salts with sodium borohydride, can be adopted; hydrogen is also generated *in situ* from a standardised solution of sodium borohydride which is introduced into a specially designed apparatus with a valve at the end of a burette so that reagent is added only as long as the sample is taking up hydrogen. The amount of hydrogen can be calculated accurately from the volume of standard sodium borohydride solution added. A modification of this procedure, that has been used for the determination of unsaturation in fats and oils, has been described.[350]

F. Location of Double Bonds in Fatty Acid Chains

The positions of double bonds in alkyl chains are generally determined by oxidative fission across the double bond followed by gas chromatographic identification of the products. Lipid analysts have largely accepted two procedures as suitable for the purpose; oxidation with permanganate–periodate reagent[122,123] (frequently termed von Rudloff oxidation) and ozonolysis followed by reductive cleavage of the ozonide. The former method yields mono- and dibasic acids as the products while the latter give aldehydes and aldehyde esters. Both procedures have been reviewed in some detail by Privett.[405]

1. *Permanganate–periodate oxidation*

In this procedure, the methyl ester of the unsaturated fatty acid in *tert*-butanol solution is oxidised by a solution containing a small amount of potassium permanganate together with a larger amount of sodium metaperiodate which continuously regenerates the permanganate as it is used up while the whole is buffered by a solution of potassium carbonate. When reaction is complete, the solution is acidified, excess oxidant is destroyed by addition of sodium bisulphite or better by passing sulphur dioxide into the solution and the products are extracted thoroughly with diethyl ether before being methylated for GLC analysis. Very little overoxidation occurs with this reagent, but it is not easy to achieve quantitative isolation of short-chain mono- and dibasic acids or half-esters of these for GLC analysis. The problem has been partially resolved by injecting the free short-chain fatty acids directly on to a GLC column of Porapak Q,[182] although carbowax 20 M–terephthalic acid might have been a better choice of liquid phase, or by pyrolysing the tetramethylammonium salts of the acids, thus converting them to methyl esters, in the heated injection port of a gas chromatograph.[123] Where the compound to be oxidised consists of more than one positional isomer, only the longer-chain fragments are obtained in reproducible yields, and wherever possible, these alone should be considered when determining the amounts of individual positional isomers.[195] The recommended procedure is as follows (only the highest purity reagents should be used).

"A stock oxidant solution is made up of sodium metaperiodate (20·86 g) and potassium permanganate (0·395 g) in water (1 l). This solution (1 ml) together with potassium carbonate solution (1 ml; 2·5 g/l) is added to the monoenoic ester (1 mg) in *tert*-butanol (1 ml) in a test-tube and the mixture is shaken thoroughly at room temperature for 1 hr. At the end of this time, the solution is acidified with one drop of concentrated sulphuric acid and excess oxidant is destroyed by passing sulphur dioxide into the solution which is then extracted thoroughly with diethyl ether (3 × 4 ml). The organic layer is dried over sodium sulphate before the solvent is removed on a rotary evaporator or in a stream of nitrogen at room temperature. The products are methylated for GLC analysis preferably by reaction with diazomethane freshly prepared by the procedure of Schlenk and Gellerman[461] (see Chapter 4)."

The procedure can be scaled up for polyunsaturated fatty acids but the proportion of water to *tert*-butanol should be kept as close to 2:1 (v/v) as possible. Malonic acid, formed by oxidation of methylene-interrupted double bonds, is oxidised further and is not detected. Ambiguity may result when polyunsaturated fatty acids with more than one methylene group between the double bonds are oxidised as it is then not possible to state which dibasic fragment contained the original carboxyl group. However, this difficulty can be resolved by reducing the carboxyl group to an alcohol prior to the analysis[170] or by repeating the analysis on the compound after extending its aliphatic chain by one carbon atom.[577]

2. *Ozonolysis and reductive cleavage*

Ozone attacks olefins rapidly and quantitatively to form ozonides with no over-oxidation and very few side reactions if the reaction is carried out at low temperature in an inert solvent such as pentane. The ozonide can be cleaved reductively by catalytic reduction with Lindlar's catalyst and hydrogen or chemically by a number of reagents of which the mildest is probably dimethyl sulphide in methanol.[419] The following procedure is generally satisfactory.

"A solution of ozone in pentane is prepared by bubbling oxygen containing ozone through purified pentane at −70°C until a blue colour is obtained. The unsaturated ester (1 mg) is dissolved in pentane (1 ml) and cooled to 0°C when the ozone solution is added dropwise until the blue colour persists. After 1 min, the reagents are removed in a stream of nitrogen. Methanol (0·3 ml) then dimethhly sulphide (0·5 ml), both precooled to −70°C, are added and the mixture is maintained at this temperature for 20 min before it is allowed to warm up to room temperature. The reagents are removed in a gentle stream of nitrogen and the products are dissolved in pentane or hexane for injection into the gas chromatograph for analysis."

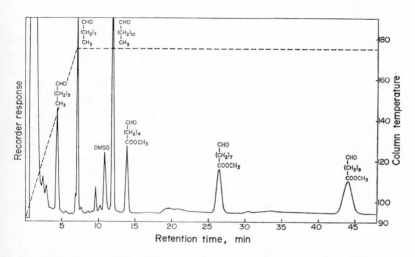

Fig. 5.8. GLC recorder trace of the ozonolysis products of a mixture of methyl petroselinate, methyl oleate and methyl vaccenate (EGS was the stationary phase)[419] (reproduced by kind permission of D. G. Cornwell and the *Journal of Lipid Research*).

With larger amounts of ester, it may be necessary to add more methanol to the reaction mixture to assist the reaction. There may also be difficulties in recovering quantitatively, short-chain aldehydes or aldehyde esters for analysis. Dimethyl sulphoxide is the other product of the reaction but this does not interfere with the GLC analysis.

The products of the two types of reaction can be identified on a number of gas chromatographic columns; difunctional compounds are more rapidly eluted on silicone liquid phases but are less likely to be confused with monofunctional compounds on polyester phases. With both types of liquid phase, temperature programmed analysis is generally necessary if all the fragments must be determined. Dibasic standards are readily available for comparison but aldehyde esters are not and it may be necessary to make up a suitable standard by ozonolysis of monoenoic esters of known structure that are commercially available, e.g. methyl petroselinate, oleate and vaccenate. A GLC trace of the ozonolysis products of such a mixture on an EGS column is illustrated in Fig. 5.8.

3. *Oxidation after partial reduction of polyunsaturated acids*

The procedures described above give optimum results with mono- and dienoic fatty acids. When the structures of polyunsaturated fatty acids must be determined, especially when they contain double bonds of both *cis*- and *trans*-configuration or are part of conjugated double bond systems, more positive identification can be obtained by partially reducing the compounds prior to oxidation.[411] Hydrazine, a reagent which does not cause any double bond migration or stereomutation, is used for the purpose under conditions such that a high proportion of monoenoic compounds are formed; isomers are then found with double bonds in each of the positions in which they were present in the original polyunsaturated fatty acid. *cis*- and *trans*-Monoenes are then separated by silver nitrate TLC (see above) and their structures separately determined by one of the above methods so that the original compound is fully identified. Hydrazine reduction, which is better performed on the free acid than on the ester, is carried out as follows.

"The free fatty acid is heated in air at 35°C with 100 volumes of 10 per cent hydrazine hydrate in methanol for a predetermined time (1·5–2 hr), found by trial and error with a standard polyunsaturated acid, such that there is an approximately 50 per cent yield of monoenes. Excess methanolic hydrogen chloride (6 per cent, w/w) is then added to stop the reaction and the mixture is refluxed for 2 hr to convert the acids to the methyl esters which are recovered for further study as described earlier."

A number of variations on the procedure exist.[405]

Ozonolysis techniques have the advantages that over-oxidation and spurious by-product formation is negligible and recovery of short-chain fragments is less of a problem, but it is necessary to have equipment to generate ozone although this is probably a worthwhile investment if many samples must be analysed. Ozonolysis can also be used when other functional groups such as free hydroxyl or epoxyl are present. The permanganate–periodate procedure, on the other hand, uses readily available inexpensive chemicals and over-oxidation is minimal. It is especially valuable when only the occasional sample must be analysed.

G. Location of Other Functional Groups in Fatty Acids

Spectroscopic aids to the recognition and location of functional groups in fatty acid chains are discussed above.

Isolated triple bonds in fatty acids are difficult to recognise as they do not exhibit particularly distinctive features when examined by any of the spectroscopic techniques, but a specific TLC spray consisting of 4-(4'-nitrobenzyl)-pyridine (5 per cent) in acetone, which gives a violet colour with acetylenes, has been described.[475] The position of the triple bond can be located by mass spectrometry of the keto-compound prepared by acid hydrolysis of the mercuric acetate adduct (see above) and they are readily cleaved by the permanganate-periodate reagent. Ozone will also react with triple bonds, although more slowly than with double bonds, and the products of reductive cleavage of the ozonide are mono- and dibasic acids rather than aldehydes and aldehyde esters so that double and triple bonds in a single fatty acid can be differentiated by this technique.[403] In addition, double bonds are hydroxylated by peracids while triple bonds remain unchanged.[404]

Allenic groups have distinctive IR and NMR spectra and all the natural fatty acids containing this functional group exhibit a marked optical activity. The position of the group in the fatty acid chain can be determined by partial reduction and oxidation of the monoene fragments as described above.[37]

A number of spectroscopic procedures can be used to detect and locate oxygenated functional groups, such as keto, hydroxyl or epoxyl, in fatty acid chains and these were described earlier. In addition, several chemical techniques are available;[414] for example, the presence of hydroxyl groups can be confirmed by GLC analysis before and after the preparation of volatile derivatives as discussed above. The first step in identifying an acid of this type is to determine its chain length. The acid is first hydrogenated to eliminate any multiple bonds before free hydroxyl or epoxy groups are converted to iodides by the action of iodine and red phosphorus. The iodine atom in the aliphatic chain is then removed by hydrogenolysis with zinc and hydrochloric acid in methanol when the resulting saturated straight-chain compound is identified by gas chromatography[343] (see Fig. 5.9a). Keto groups should be reduced to hydroxyl groups by the action of sodium borohydride[500] prior to analysis by the above procedure. The hydroxyl group can be located in the fatty acid chain unequivocally by the reaction illustrated in Fig. 5.9b. It is first oxidised with chromic oxide in acetic acid to a keto group from which the oxime derivative is prepared. This is heated with concentrated sulphuric acid so that it undergoes a Beckman

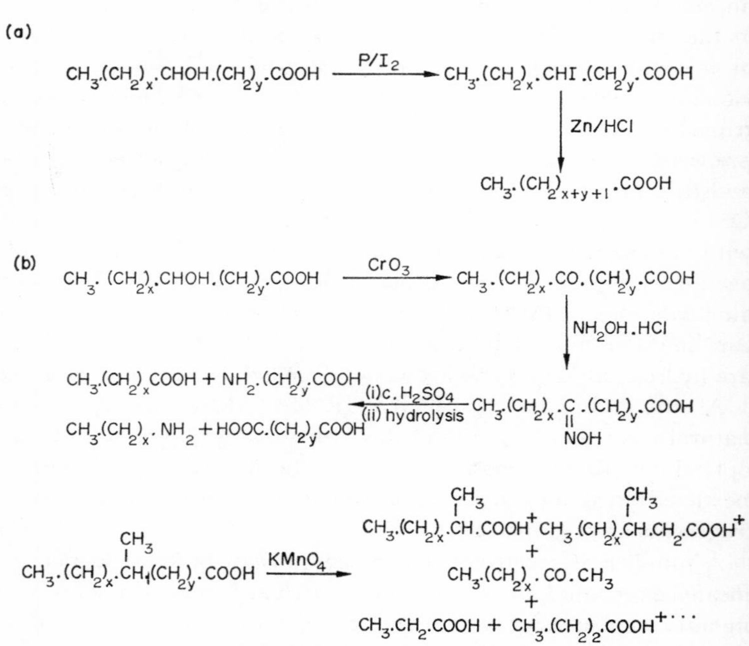

(a)

$$CH_3.(CH_2)_x.CHOH.(CH_2)_y.COOH \xrightarrow{P/I_2} CH_3.(CH_2)_x.CHI.(CH_2)_y.COOH$$

$$\downarrow Zn/HCl$$

$$CH_3.(CH_2)_{x+y+1}.COOH$$

(b)

$$CH_3.(CH_2)_x.CHOH.(CH_2)_y.COOH \xrightarrow{CrO_3} CH_3.(CH_2)_x.CO.(CH_2)_y.COOH$$

$$\downarrow NH_2OH.HCl$$

$$CH_3.(CH_2)_x.COOH + NH_2.(CH_2)_y.COOH$$
$$\xleftarrow[\text{(ii) hydrolysis}]{\text{(i)c. }H_2SO_4}$$
$$CH_3.(CH_2)_x.NH_2 + HOOC.(CH_2)_y.COOH$$

$$CH_3.(CH_2)_x.\underset{\underset{NOH}{\|}}{C}.(CH_2)_y.COOH$$

(c)

$$CH_3.(CH_2)_x.\overset{\overset{CH_3}{|}}{CH}(CH_2)_y.COOH \xrightarrow{KMnO_4}$$

$$CH_3.(CH_2)_x.\overset{\overset{CH_3}{|}}{CH}.COOH^+ \ CH_3.(CH_2)_x.\overset{\overset{CH_3}{|}}{CH}.CH_2.COOH^+$$
$$+$$
$$CH_3.(CH_2)_x.CO.CH_3$$
$$+$$
$$CH_3.CH_2.COOH + CH_3.(CH_2)_2.COOH^{+\cdots}$$

Fig. 5.9. Location of functional groups in fatty acids by chemical means. a, Determination of chain length in a hydroxy fatty acid. b, Location of hydroxyl- or keto-groups in fatty acids. c, Location of methyl branches in fatty acids.

rearrangement reaction with formation of isomeric amides, which are hydrolysed and the acidic products isolated, methylated and identified by gas chromatography[93] (it is only necessary to identify one of the products to be able to completely identify a monohydroxy acid). Oxirane rings are cleaved directly with periodic acid in halogenated solvents[46] or in diethyl ether[294] and the position of the ring is established by analysis of the products.

The chemistry of cyclopropane and cyclopropene ring-containing fatty acids has been reviewed comprehensively elsewhere.[89] The presence of such functional groups can be detected by a number of spectroscopic procedures and also by various chemical techniques. For example, cyclopropene rings give a pink coloration with carbon disulphide (the Halphen test) and a brown coloration with silver nitrate. The ring is disrupted by ozonolysis or permanganate–periodate oxidation and a β-diketo compound is formed which can be identified by mass spectrometry[222] or this compound can be subjected to more vigorous oxidation such that the molecule is cleaved between the keto groups when the resulting acidic fragments can be recovered and identified. Cyclopropene rings can also be hydrogenated and their positions determined by the methods used for natural cyclopropane fatty acids. The latter behave as normal saturated fatty acids on silver nitrate plates but react with bromine so can be distinguished by this means.[61] They form methoxy-derivatives with boron trifluoride–methanol reagent[347,348] and they are converted to methyl-branched fatty acids by vigorous catalytic hydrogenation[321] and both types of derivative can be characterised by mass spectrometry. The latter derivatives can also be characterised by the chemical technique described below.

Methyl branches in fatty acids are not easily located by chemical means as such compounds are comparatively inert to most chemical reagents.[2] Their positions can be determined chemically by oxidising the fatty acids vigorously with acidic potassium permanganate. Homologous series of normal and branched-chain acids are formed together with a neutral keto compound that identifies the point of branching. Both the acidic and keto products can be isolated and identified by gas chromatography[367,378] (see Fig. 5.9c). These and related procedures have also been reviewed by Polgar.[399]

H. Physical Characterisation of Fatty Acids

Melting points of long-chain saturated or oxygenated fatty acids are often sufficiently distinct to be useful in their characterisation; they are commonly determined on the free acids rather than on the methyl esters as the melting points of the former tend to be significantly higher. Unsaturated acids are in general low melting and while a melting-point apparatus equipped with a cold stage may be used to characterise monoenes (*trans*-isomers have higher melting points than the corresponding *cis*-compounds) and dienes, the melting points of poly-unsaturated fatty acids are in general too low to be measured with confidence as traces of impurities such as autoxidised materials can greatly affect this property. An apparatus for semi-micro crystallisations of fatty acids at low temperatures for subsequent melting point determination has been described by Privett.[406] *p*-Phenylazophenacyl esters of polyunsaturated fatty acids, which have comparatively high melting points, have been suggested as suitable derivatives for characterisation purposes,[557] but have not been widely adopted.

Measurement of *critical solution temperatures* may be an acceptable alternative to melting points when the latter cannot conveniently be measured. The method utilises the property that two liquids, which are almost immiscible at low temperatures, become more soluble in each other as the temperature is raised. The temperature at which they become soluble in all proportions is known as the upper critical solution temperature. Nitromethane or acetonitrile have been used as solvents to characterise a wide range of polyunsaturated fatty acids and related compounds in the systematic studies of Schmid *et al.*,[468] but the technique has not been as widely used as it deserves. Other physical methods of characterising low-melting fatty acids have been reviewed by Holman *et al.*[217] Chromatographic data (e.g. equivalent chain length values) are now frequently accepted as physical constants.

I. Preparation of Large Quantities of Pure Fatty Acids

Although a number of methods are available for the preparation of pure fatty acids in large quantities (50 g or more) as fatty acid standards or for biochemical or nutritional studies, many procedures require special equipment likely to be available in few laboratories. For example, highly efficient fractional distillation apparatus is used by most commercial suppliers in the preparation of fatty acids. Several less sophisticated methods have been developed, however, that can be used to advantage by lipid analysts. The key to success frequently lies in the choice of suitable starting materials.

1. *Saturated fatty acids*

Coconut oil is commonly used as a source of C_8 to C_{14} fatty acids and beef tallow or hydrogenated fats for longer chain esters. Certain seed oils are also useful sources of specific acids; for example, bayberry tallow may contain up to 60 per cent myristic acid.

Concentrates of specific shorter chain acids are obtained by high vacuum distillation of coconut oil methyl esters with comparatively simple fractional distillation columns, but more expensive equipment is necessary to obtain highly pure esters in quantity by this technique.

The liquid–liquid chromatography system of Privett and Nickell[406,410] described above (section C4) is a useful alternative. A column (5·5 × 125 cm) containing 1000 g of solid support has been used to purify up to 15 g of esters and an even larger column to purify up to 50 g of esters in a single run. The columns can be used repeatedly.

However, most of the common saturated fatty acids are now obtainable in a pure form comparatively cheaply from commercial suppliers so that simple laboratory methods of preparing them are becoming less important. Odd-numbered fatty acids can be prepared synthetically, if necessary, by chain-elongation of the more common even-numbered acids.

2. *Polyunsaturated fatty acids*

(i) *Sources*

Sources of polyunsaturated fatty acids and methods of preparing them have been described in an excellent review by Privett.[406] Again it is advisable to start with some natural material rich in the acid required.

For example, olive oil is an excellent source of oleic acid, safflower oil of linoleic acid, linseed oil of linolenic acid, pig liver lipids of arachidonic acid and codliver oil of penta- and hexaenoic acids. Usually it is necessary to obtain a fraction rich in the particular component required before proceeding to more exacting fractionation procedures.

(ii) *Low temperature crystallisation*

Fractions enriched in polyunsaturated fatty acids are readily obtained by low temperature crystallisation, a technique that demands little in the way of expensive equipment. Most methods employ acetone for crystallisation of samples in the form of methyl esters or acetone, diethyl ether or hexane for the free acids, which are generally preferred for the purpose, at temperatures down to $-70°C$, a temperature readily attained with solid carbon dioxide as refrigerant. Brown and Kolb[69] have described a number of useful procedures.

Three fractions enriched in saturated, monoenoic and polyenoic fatty acids respectively are easily obtained. The free acids are taken down to the lowest working temperature first (about $-50°C$ generally) at a concentration of 1 g per 10 ml of acetone and are kept at this temperature for up to 5 hr with gentle swirling until equilibrium is reached. The solution is then filtered through a Buchner funnel cooled to just below the solution temperature and the crystals washed with a small amount of cold solvent. The material in solution consists mainly of polyunsaturated fatty acids although it is always contaminated by small amounts of saturated and monoenoic components held in solution by the polyunsaturated compounds. If the volume of solvent is reduced until a 10 per cent solution is again attained, the process can be repeated and improved separations obtained. The crystals may be redissolved in fresh solvent and further fractionated at slightly higher temperatures into components enriched in monoenoic and then in saturated fatty acids.[107] With care some unexpected separations can be achieved; for example, petroselinic acid has been separated from oleic acid by this technique.[409]

The main disadvantages of the method are that the fractions are not entirely pure and that the solutions must be held at the low temperatures for very long periods before equilibrium is attained. In the method's favour, large quantities of fatty acids can be processed and very little harm can come to polyunsaturated acids at the low temperatures employed.

(iii) *Urea adduct formation*

Pure urea, when permitted to crystallise in the presence of certain long-chain aliphatic compounds, forms hexagonal crystals with a channel into which the aliphatic compounds may fit and be removed from solution, provided they do not contain functional groups that increase their bulk. Such crystals are known as urea inclusion complexes. Saturated straight-chain acids (as the methyl ester derivatives) form complexes readily. The double bonds of unsaturated fatty acids increase their bulk so that monoenoic fatty acids do not form complexes easily but tend to form them more readily than dienes which in turn form complexes more readily than compounds with three or more double bonds. Fatty acids with *trans*-double bonds form complexes before the analogous compounds with *cis*-double bonds. Unfortunately, the separations are complicated by the fact that shorter chain-length compounds do not complex as readily as those of longer chain-length and methyl oleate for example is adducted with approximately the same facility as methyl laurate.[236] For this reason, the procedure has never been developed as an analytical technique. The following method can be applied to obtain a concentrate of polyunsaturated fatty acids in the form of methyl esters from a natural mixture.

"The esters (100 g) are dissolved in methanol (1 l) to which urea (200 g) is added. The mixture is warmed until all the urea has dissolved when the solution is allowed to cool to room temperature with occasional swirling. After a minimum of 4 hr, the material is filtered through a Buchner funnel to remove the urea complexes which are washed twice with 25 ml portions of methanol saturated with urea. The solution, which is greatly enriched in polyunsaturated esters, is concentrated to a volume of about 200 ml on a rotary evaporator then poured into 1 per cent hydrochloric acid (600 ml) and extracted alternately with hexane (500 ml) and diethyl ether (500 ml). The combined organic layers are washed twice with water (50 ml

F

portions) and dried over anhydrous sodium sulphate before the solvent is removed under reduced pressure."

The procedure can be scaled up or down considerably. For example, a GLC recorder trace of material obtained in this way from 0·25 g of the standard mixture described above (see Fig. 5 1) is illustrated in Fig. 5.10. The fraction obtained, which was 20 per cent of the original

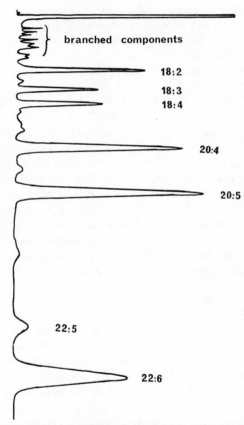

Fig. 5.10. GLC recorder trace (the EGSS-X column of Table 5.1 and Fig. 5.1) of methyl esters that did not form urea adducts from the natural mixture illustrated in Fig. 5.1.

esters, is enriched in the polyunsaturated components but also contains some branched-chain esters, not readily apparent earlier, and some shorter-chain-length esters. *iso*-Branched esters complex more readily than the corresponding *anteiso*-compounds so some change in the ratio of these may occur.

The adducted esters can also be recovered, if necessary, by breaking up the complexes with water and extracting the esters into hexane or diethyl ether.

Fractions enriched in particular fatty acids such as linoleic or linolenic can also be obtained from suitable starting materials although it may be necessary to vary the amount of urea added. The procedure has also been used to obtain concentrates of branched-chain and cyclic esters from natural mixtures and also to separate ω-hydroxy acids (after acetylation to increase their bulk) free of hydroxy acids with the substituent elsewhere in the chain.[94] Attempts to use urea in thin-layer adsorbents or in columns on an analytical scale have not been entirely successful, but the following simplified procedure is of value when small amounts only of esters are available.[304]

"The methyl esters (100 mg) are dissolved in hexane (4 ml) and urea (1·5 g) moistened with methanol (15 drops) added. After standing overnight, the solid is filtered off and thoroughly washed with hexane; the washings and the hexane filtrate are combined, washed with water, dried over anhydrous sodium sulphate and evaporated yielding the branched-chain or polyunsaturated fraction. The straight-chain fraction can be recovered by dissolving the urea in water, extracting with diethyl ether, drying the solvent layer and evaporating as before."

The main advantages of methods using urea are that large quantities of esters can be separated and that with care there is little chance for harm to come to polyunsaturated esters. Urea fractionation procedures have been reviewed comprehensively by Schlenk.[460]

(iv) *High capacity adsorption chromatography*

When the starting material is plentiful so that efficiency of separation is not important, adsorption chromatography on large columns of silicic acid can be used to purify fatty acids. Generally in conventional adsorption chromatography only those components held least strongly are eluted in a pure fraction but concentrates of the more strongly held components can be obtained by using silicic acid for displacement chromatography, i.e. those components held most strongly serve as internal displacers for those held less strongly. Privett *et al.*[368,406] have described an elegant application of this method for the preparation of 100 g quantities of methyl linolenate (>99 per cent pure).

(v) *Separation as mercuric acetate adducts*

Concentrates rich in polyunsaturated fatty acids can be obtained from suitable starting materials by preparing the mercuric acetate adducts (see Chapter 4 for practical details of the preparation of adducts and regeneration of the original double bonds) of methyl esters and partitioning them between methanol and pentane; the methanol retains the adducts of the more unsaturated esters which can be regenerated unchanged. Methyl linoleate of 95 per cent purity and methyl linolenate of 90 per cent purity can be produced on the 50–100 g scale by this method from the esters of safflower and linseed oils respectively.[515] The author has prepared methyl linoleate of greater than 99 per cent purity simply by partitioning the adducts of safflower oil esters between methanol–water (95:5, v/v) and hexane in separating funnels (unpublished work) and others[579] have obtained methyl linolenate of similar purity by partitioning the adducts of linseed oil esters between methanol–water (90:10, v/v) and diethyl ether in a liquid–liquid continuous extractor.

The method deserves much wider recognition. Although mercuric acetate is not cheap, large quantities of high purity unsaturated esters can be obtained in a relatively short time. If these compounds are to be used in biochemical studies, however, particularly in nutritional experiments, great care must be exercised to remove all traces of mercury. This aspect of the process requires further study.

(vi) *Fractional distillation*

Because of the highly specialised nature of the equipment used, this procedure will not be discussed in detail here. It must be noted, however, that it is the only method available for the preparation of pure fatty acids with four or more double bonds from natural sources in quantity. Again the excellent review by Privett[406] discusses such procedures at length.

CHAPTER 6

The Analysis of Simple Lipid Classes

LIPID extracts from animal tissues commonly contain triglycerides as the predominant simple lipid class together with cholesteryl esters, cholesterol, partial glycerides and free fatty acids in addition to complex lipids, which in this chapter are considered as a single class. As well as these more common simple lipids, trace amounts of hydrocarbons, methyl esters, wax esters and glyceryl ethers and vinyl ethers (neutral plasmalogens) may be found and, on occasion, one or more of these may assume major proportions. Plant lipid extracts may contain similar simple lipids except that sterols other than cholesterol are likely to be present. Most of these lipid classes can be separated from each other by adsorption chromatography on thin layers or on columns. Where the fatty acid constituents of lipids contain polar functional groups in the alkyl chain, however, simpler molecular species of these will be separated by adsorption chromatography and the elution pattern may be rather complicated; separations of this kind are discussed further in Chapter 8 and are not considered here. Precautions must always be taken to minimise the effects of autoxidation (see Chapter 3).

A. Chromatographic Separation of the Common Simple Lipid Classes

1. *Thin-layer chromatography*

(i) *Solvent systems*

A variety of solvent systems have been used to separate simple lipids by TLC on silica gel G in a single dimension. Those used most frequently contain hexane, diethyl ether and acetic (or formic) acid in various proportions; for example, with these solvents in the ratio 80:20:2 (by vol.), the separation illustrated in Fig. 6.1 (plate A) is achieved in which most of the common simple lipids are separated—cholesteryl esters near the solvent front followed by triglycerides, free fatty acids, cholesterol and diglycerides (α,α'- and α,β-), monoglycerides and phospholipids. If diglycerides are important components of the sample, more distinct separations can be achieved with a solvent system consisting of benzene–diethyl ether–ethyl acetate–acetic acid (80:10:10:0·2 by vol.)[520] as illustrated in Fig. 6.1 (plate B), but it should be recognised that considerable isomerisation of diglycerides occurs on TLC adsorbents (see below). Unfortunately, slight changes in the amount of acetic acid in the mixture can drastically alter the nature of the separation. Other useful TLC systems in which plates are developed twice in the same direction with different solvent mixtures have been described by others.[139, 493] As little as 0·5 μg can be applied as a spot to a plate if the lipids are detected by charring techniques. For preparative purposes, 20 mg of lipid may be applied with ease as a band on a 20×20 cm plate coated with a layer 0·5 mm thick of silica gel and in some circumstances, for example when the lipids consist largely

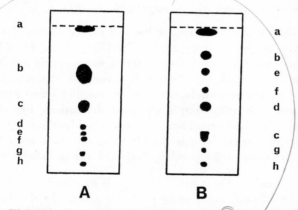

FIG. 6.1. TLC separations of simple lipids on silica gel G layers. a, cholesteryl esters; b, triglycerides; c, free fatty acids; d, cholesterol; e, α,α'-diglycerides; f, α,β-diglycerides; g, monoglycerides; h, complex lipids. Developing solvents: plate A, hexane–diethyl ether–formic acid (80:20:2 by vol.); plate B, benzene–diethyl ether–ethyl acetate–acetic acid (80:10:10:0·2 by vol.).[520]

of triglycerides, up to 50 mg of lipid can be chromatographed on a single plate in this way. Simple lipids should be applied to the plate in a solvent such as chloroform that does not evaporate too quickly depositing material on the outside of the syringe.

As cholesteryl esters migrate very close to the solvent front with these solvent systems where any lipid-like contaminants which may be in the silica gel will also appear, it is sometimes recommended that plates be developed first right to the top in a solvent system of chloroform–methanol (4:1, v/v) then dried immediately before the sample is applied so that any impurities in the adsorbent will lie ahead of the earliest running component of the mixture. Although this treatment is quite effective, the retention characteristics of the adsorbent may also be altered significantly through deactivation.

Triglycerides (and other lipids) that contain fatty acids differing widely in chain length such as are found in milk fats will migrate as a more diffuse band on TLC adsorbents as they are partially separated into molecular species that contain only long chain ($>C_{16}$) and those that contain short chain components (slower running).

(ii) *Detection and identification of simple lipids on TLC plates*

Lipid classes can be detected by any of the destructive or non-destructive non-specific reagents described in Chapter 3 and identified by their migration characteristics relative to authentic standards that are chromatographed simultaneously alongside the samples under investigation. Unfortunately, there is no specific spray reagent that can be used to identify simple glyceride derivatives, but cholesteryl esters and free cholesterol can be positively identified by means of various spray reagents (including charring reagents, see Chapter 3) of which the most useful is acidic ferric chloride solution.[314]

"Ferric chloride ($FeCl_3.6H_2O$, 50 mg) is dissolved in water (90 ml) with acetic (5 ml) and sulphuric (5 ml) acids. The developed TLC plate is sprayed with the reagent then heated at 100°C for 2–3 min when the presence of cholesterol and cholesteryl esters is indicated by the appearance of a red-violet colour. The colour for cholesterol appears slightly before that of its ester."

Free fatty acids can be identified by means of a specific sequence of spray reagents,[124] although the chemical reactions involved are not known.

"The developed plate is sprayed in turn with a 2′,7′-dichlorofluorescein spray (see Chapter 3), a solution of 1 per cent aluminium chloride in ethanol and finally with 1 per cent aqueous ferric chloride, warming the plate briefly after each spray. Free fatty acids give a rose-violet colouration."

Esterified fatty acids can also be recognised by a specific spray sequence.[488, 580]

"Two reagents are necessary. (a) Hydroxylamine hydrochloride (10 g) is dissolved in water (25 ml) and ethanol (100 ml) and added to a saturated aqueous solution of sodium hydroxide (26 ml) and ethanol (200 ml). The mixture is filtered before use to remove any sodium chloride precipitated. (b) Ferric chloride ($FeCl_3.6H_2O$; 10 g) and concentrated hydrochloric acid (20 ml) are ground together in a mortar then are shaken with diethyl ether (300 ml). The developed plate is sprayed with the first reagent, dried then sprayed with the second reagent when the esterified lipids appear as purple spots on a yellow background."

(iii) *Estimation of simple lipids*

Three basic types of procedure may be used to estimate simple lipids separated by TLC. In the first approach, all the lipids are charred by oxidising agents on or after elution from the plates and the amount of carbon formed is determined by photodensitometry; in the second, a dye is sprayed on the plate and the lipids are determined by fluorometry; while in the third approach, lipids are detected and recovered from the adsorbent and are estimated individually by methods that are appropriate to each lipid class. Procedures of the first type are rapid but are destructive to samples and are of limited accuracy; those of the third type are much more time-consuming but are capable of greater accuracy and may yield more information.

The principles of procedures that involve charring followed by photodensitometry of the carbon formed have been discussed briefly in

Chapter 3. All lipids can be estimated in a single analysis when they are charred directly on a TLC plate, but similar standards must be run simultaneously and large blank values may be obtained that limit the accuracy of the procedure. It is none the less well suited to the analysis of large numbers of similar samples when the expensive specialised equipment, which is necessary, is available. Simple lipids can also be eluted from the adsorbent with chloroform or chloroform–methanol (90:10, v/v) for charring and photodensitometry. Such procedures are generally capable of greater accuracy as calibration is easier and the Beer–Lambert laws are more likely to hold when substances are in solution in the absence of solid material. In addition, the absorbance of the solutions can be measured on a general purpose laboratory UV spectrophotometer rather than with a specialised scanning photo-densitometer. In the most commonly used procedures of this type, the lipids are oxidised with potassium dichromate solution and the reduction in absorbance of the dichromate ion at 350 nm is measured (see Chapter 3)[19, 139, 489] or the lipids can be charred at 200°C with concentrated sulphuric acid and the absorbance at 375 nm measured.[339] It is necessary to have a separate calibration curve, obtained with a suitable standard, for each lipid class.

Fluorometric methods have yet to be applied widely for lipid analysis, but hold great promise for the future. They are discussed in Chapter 3.

Cholesterol and cholesteryl esters may be estimated by the methods described below (section B). The amounts of tri-, di- and monoglycer-ides can be determined by quantitative IR spectrophotometry after they have been eluted from TLC plates.[141] After the removal of solvent, the lipid classes are dissolved in a known volume of carbon tetrachloride and the absorbance due to the characteristic frequency of the carbonyl function at 5·85 μm is measured. The amount of each component present is then obtained by reference to calibration curves, which must be regularly checked, prepared from tri-, di- and mono-olein. The hydroxamic acid colorimetric reaction for acyl-ester groups is also used to estimate simple glycerides (see section B below).

Free fatty acids can be estimated titrimetrically as follows.

"After removal of the solvent used to elute the acids from the TLC plate, the acids are taken up in 95 per cent ethanol–diethyl ether

(1:1 by vol.; 5 ml) and cresol red (0·1 ml; 0·005 per cent) is added as indicator. 0·005 N sodium hydroxide in 70 per cent ethanol is then used to titrate the acids while a stream of nitrogen is bubbled through the mixture to stir the reagents and to exclude atmospheric carbon dioxide.''

The spray reagents used to detect lipids may interfere with colorimetric or spectrophotometric determinations if they are applied too liberally to the plates.

Complex lipids considered as a single class can be estimated by phosphorus determination (assuming that non-phosphorus containing glycolipids are not major components of the sample) by the procedure described in Chapter 7. The approximate amount of phospholipid is obtained by multiplying the weight of phosphorus found in the sample by the factor 25.

If, in addition to estimating the amount of each lipid class in a natural mixture, it is intended that the fatty acid composition of each be determined, then both quantities can be obtained in a single analysis by gas chromatography of the methyl ester derivatives of the fatty acid components with an added internal standard (see Chapter 3 for a description of the underlying principle). A known amount of the standard, which should be the methyl ester of a fatty acid that does not occur in the sample to be analysed, is added to each lipid class as it is eluted from the TLC adsorbent before it is transesterified and the methyl esters subjected to GLC analysis (see Chapter 5). The fatty acid composition of the sample is in this way determined and the amount of the sample can be found by relating the total area of the fatty acid peaks to the area of the peak for the standard ester.[99] It is necessary to allow for the weight of non-fatty acid material (e.g. glycerol or cholesterol) in each lipid class by multiplying each result by appropriate arithmetic factors calculated by dividing the molecular weight of the heptadecanoic acid derivative (assuming 17:0 is the internal standard) of the lipid class by the molecular weight of methyl heptadecanoate. These factors are listed in Table 6.1.[99] Alternatively, an average molecular weight for each lipid class can be calculated from the fatty acid compositions so other factors can be derived from this.

TABLE 6.1. CORRECTION FACTORS TO CONVERT THE TOTAL AMOUNT OF
FATTY ACIDS IN A LIPID CLASS (DETERMINED BY GLC) TO WEIGHT OF LIPID[99]

Lipid class	Factor
Cholesteryl esters	2·246
Triglycerides	0·995
Diglycerides	1·049
Monoglycerides	1·211
Free fatty acids	0·951
Phospholipids	1·371*

* As an approximation; it is assumed that the phospholipids consist of phosphatidyl choline.

Methyl heptadecanoate or pentadecanoate are chosen most frequently as internal standards and are usually added in methanol solution (approx. 0·25 mg/ml). Suitable transesterification procedures have been described in Chapter 4; methanolic hydrogen chloride can be used to esterify all lipid classes (remembering that an inert solvent must be added to effect solution of triglycerides and cholesteryl esters) or free fatty acids alone may be esterified by this reagent and the other lipid classes by the more rapid and milder alkaline transesterification procedures. Losses can be minimised by carrying out the reaction in the vessel used to collect the eluant so that no transfers are carried out before all the fatty acids are methylated or better by transesterifying lipids without prior elution from the adsorbent.[90] In the latter circumstance, the solution of the internal standard, the transesterifying reagent and a solvent (if necessary) are added together to the lipid on the adsorbent in a test-tube (see Chapter 4 for further details). With these precautions, the method is at least as accurate as any other in current use. Free cholesterol must of course be determined separately.

2. *Column chromatography*

Larger amounts of simple lipids can be separated preparatively by column chromatography with silicic acid, acid-washed Florisil or Florisil as adsorbents (see Chapter 3) eluted in a stepwise sequence with hexane containing increasing proportions of diethyl ether. The best resolutions of lipid classes are obtained with fine mesh silicic acid[204] which yields separations of the kind illustrated in Fig. 6.2. Hydrocarbons, cholesteryl esters and triglycerides are all well separated (though not illustrated free fatty acids elute between triglycerides and free cholesterol), cholesterol and diglycerides overlap slightly, but monoglycerides in turn are well separated. The precise amount of diethyl ether that must be added to the hexane to elute each lipid class will vary from batch to batch of the adsorbent so will not always be the same as that given in Fig. 6.2 and must be determined by experiment. Complex lipids can be recovered as a single class by elution with methanol or individual complex lipids can be separated by eluting with chloroform containing increasing proportions of methanol (see Chapter 7). Acid-washed Florisil has similar adsorptive properties to silicic acid, but much larger amounts of material may be chromatographed at much faster flow rates, although some slight loss of resolution is inevitable.[74]

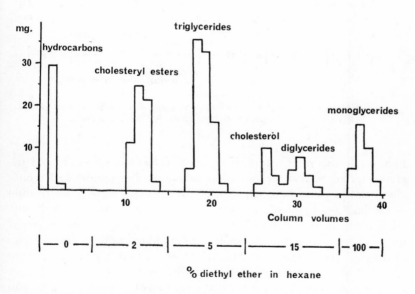

Fig. 6.2. Chromatography of simple lipids on a column (1·5 × 20 cm) containing silicic acid (30 g) eluted with hexane–diethyl ether mixtures.

Florisil itself is also a useful adsorbent with somewhat different elution characteristics to silicic acid.[73] The degree of hydration of the adsorbent is particularly critical and it may be necessary to add up to 7 per cent water to it for optimum resolution and to minimise "tailing".

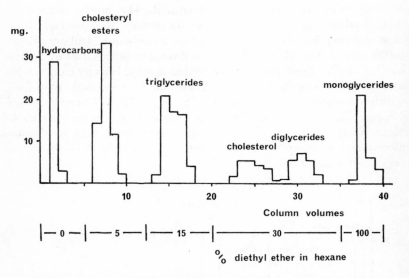

FIG. 6.3. Chromatography of simple lipids on a column ($1\cdot5 \times 20$ cm) containing Florisil (30 g) eluted with hexane–diethyl ether mixtures.

Figure 6.3 illustrates a typical separation. Cholesteryl esters and triglycerides are well separated and cholesterol, diglycerides and monoglycerides can also be resolved with care, though some isomerisation of the last two lipids occurs. Free fatty acids are eluted with diethyl ether–acetic acid (98:2 v/v), but complex lipids are not easily recovered quantitatively. When large amounts of cholesteryl esters, triglycerides or methyl esters must be isolated from natural mixtures or from synthetic preparations for further studies and complete recovery or resolution of other lipids is unnecessary, Florisil chromatography can be especially recommended as up to 1 g of lipid can be separated rapidly on columns containing 20 g of adsorbent. Though the adsorbent is commonly thrown away after use, Florisil and acid-washed Florisil can apparently be regenerated and re-used repeatedly by washing the columns with methanol followed by chloroform and then by hexane.[75]

Simple lipids as a class can be separated from complex lipids on silicic acid or acid-washed Florisil columns (30 mg/g of adsorbent) as chloroform (10 volumes) elutes simple lipids quantitatively while the complex lipids are recovered by elution with methanol (10 volumes).

The general procedures that are used for determining lipids eluted from columns have been discussed in Chapter 3. In addition, the specific chemical or spectroscopic procedures discussed earlier in this chapter for TLC separations may be used.

3. *Gas–liquid chromatography*

It is possible to determine plasma lipids by a rather novel application of high-temperature gas chromatography.[293] The lipids are first digested with phospholipase C (see Chapter 9) which converts lysophosphatidyl choline, phosphatidyl choline and sphingomyelin to monoglyceride, diglyceride and ceramide respectively. These are converted to volatile derivatives and injected directly into a gas chromatographic column containing a highly thermostable stationary phase. With tridecanoin as internal standard, triglycerides, cholesteryl esters, free cholesterol and the choline-containing complex lipids can be determined. The problems of GLC analysis of intact lipids of high molecular weight are discussed in detail later (Chapter 8).

4. *Separation of isomeric mono- and diglycerides*

Although isomeric diglycerides are separable by TLC on silica gel by the procedures described above, some isomerisation inevitably occurs on the plate and it will also occur with adsorbents in columns. With natural lipid mixtures from tissues, isomerisation will almost certainly have occurred in the extraction step and, although 1,3-diglycerides in particular may be separated and estimated by certain of the above procedures, they may originally have been 1,2-diglycerides in the native state. Therefore, 1,3- and 1,2-diglycerides should often be determined together as misleading results may be obtained if they are analysed separately.

Partial glycerides can be recovered from tissues with minimal isomerisation, if this is necessary, by extracting the tissues with non-alcoholic solvents, taking care not to heat extracts at any stage, although it should be recognised that other lipid constituents will not be recovered completely. In addition, partial glycerides are also formed as essential steps in chemical or enzymatic degradation of other lipids (e.g. in stereospecific analysis of triglycerides or on phospholipase C hydrolysis of phosphoglycerides; see Chapter 9) or by chemical synthesis, and in such circumstances they must be purified by chromatographic procedures in which the isomerisation permitted is minimal. The simplest method by which this can be achieved is to chromatograph the partial glycerides on TLC plates coated with silica gel G impregnated with boric acid (see Chapter 3 for details of preparation) using a solvent system of chloroform (alcohol-free)–acetone (96:4, v/v).[535] α,β- and α,α'-Diglycerides and α- and β-monoglycerides can be cleanly and safely separated by this procedure as illustrated in Fig. 6.4. Lipids are detected by means of the 2',7'-dichlorofluorescein spray and recovered from the adsorbent by elution with chloroform–acetone (9:1, v/v)

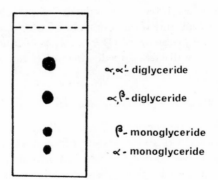

Fig. 6.4. Thin-layer separation of partial glycerides on silica gel G containing 5 per cent (w/w) boric acid. Solvent system; chloroform–acetone (96:4 v/v).[535]

which should be washed with ice water to remove any boric acid and dried rapidly over anhydrous sodium sulphate before removal of the solvent at ambient temperature. Diglycerides will isomerise slowly on standing in inert solvents or in the dry state even at low temperatures and if they are required for any synthetic or biosynthetic purpose, they should be used as soon as possible after preparation.

Up to 20 mg of partial glycerides can be purified preparatively on a 20 × 20 cm plate coated with a layer 0·5 mm thick of silica gel impregnated with 10 per cent boric acid but larger amounts can be purified by column chromatography on acid-washed Florisil impregnated with boric acid.[478]

Individual isomers of partial glycerides have distinctive NMR spectra[478] that can be used to aid their identification.

B. Analysis of the Hydrolysis Products of Simple Lipids

For unequivocal identification of simple lipids, it is necessary to isolate and determine all the hydrolysis products to ascertain the molar proportions of each in the lipid.

1. *Fatty acids*

The estimation of the total amount of fatty acids in simple lipids by means of gas chromatography with an internal standard has been discussed above. They can also be determined chemically by the ferric hydroxamate procedure.[121, 421]

"Two reagents are necessary and must be prepared immediately before use. The first is prepared by dissolving hydroxylamine hydrochloride (100 mg) in water (0·17 ml) and adding absolute alcohol (3·2 ml). To this is added aqueous sodium hydroxide solution (0·2 ml; 5 g per 10 ml) in absolute alcohol (3·2 ml). Sodium chloride formed is removed by centrifugation. The second reagent consists of ferric perchlorate (100 mg) and perchloric acid (2·6 ml) in 95 per cent ethanol (97·4 ml). The samples and a set of standards containing 0·5 to 5·0 μeq. of fatty esters are dried thoroughly in a vacuum oven at 65°C (in particular, all traces of chloroform must be removed). The basic hydroxylamine reagent (0·1 ml) is added to each sample in diethyl ether solution (3 ml) and the whole is heated at 65°C at atmospheric pressure until the ether evaporates and then in a vacuum oven until all the remaining solvent is gone. The samples are cooled, ferric perchlorate reagent (5 ml) is added and the mixture shaken and left to stand for 30 min when the absorbance of the solution at 530 nm is read against that of a reagent blank. The amount of ester in each unknown is determined from the curve obtained with the standards."

2. *Glycerol*

The chemical determination of glycerol is time-consuming and tedious and subjected to a number of errors. In the generally accepted procedure, glycerol is oxidised with periodate liberating formaldehyde which is reacted with chromotropic acid in sulphuric acid solution; the resulting coloured product is determined spectrophotometrically.[188, 575] A more sensitive procedure utilises a coupled enzyme assay. The simple glycerides are hydrolysed with tetraethylammonium hydroxide under conditions such that phosphate ester bonds are unaffected. Glycerokinase is then utilised to convert glycerol into glycerol-3-phosphate which is in turn oxidised by α-glycerophosphate dehydrogenase in the presence of NAD. The amount of NADH produced, which is measured by spectrophotometry or fluorometry, is directly proportional to the amount of glycerol present.[84, 553] The procedure may be used with as little as 0·05 μmoles of glycerol. It is possible to purchase all the necessary enzymes and reagents separately for the method, but a number of biochemical suppliers offer kits containing the required reagents in the correct proportions for the assay together with full instructions.

Gas chromatographic procedures for estimating glycerol have been reviewed by Roberts.[438] The following method is suitable for the determination of glycerol in simple glycerides; the glycerides are subjected to hydrogenolysis with lithium aluminium hydride yielding glycerol and fatty alcohols, which are acetylated for GLC analysis.[210]

"The glyceride (3–10 mg) and a known amount of a suitable methyl ester as internal standard (e.g. methyl pentadecanoate) are dissolved in dry diethyl ether (2 ml) and a solution of lithium aluminium hydride (20 mg) in dry diethyl ether (3 ml) is added in 0·1 ml portions until the boiling stops. A one volume excess of the lithium aluminium hydride solution is added and the mixture refluxed for an hour. Acetic anhydride is added dropwise with cooling to decompose excess reagent, followed by further acetic anhydride (2·5 ml) and xylene (3 ml). The ether is evaporated off and the solution refluxed for 6 hr (the reflux temperature should be at least 110°C if no ether remains) when the reagents are removed on a rotary evaporator and the products taken up in dry diethyl ether for GLC analysis."

The procedure can be scaled down considerably. Triacetin elutes just before octadecanyl acetate on an EGS column at 170°C but very much earlier (near octanyl acetate) on non-polar stationary phases such as ApL or SE-30.[230] It is necessary to determine experimentally the molar response factor of the GLC detector to triacetin relative to those of the alcohol acetates, but with care the ratio of glycerol to fatty alcohols and, therefore, to fatty acids and indeed the fatty acid composition of the original glyceride can be accurately determined in a single analysis. The analysis of fatty alcohol acetates is discussed in greater detail below.

3. *Cholesterol*

Free cholesterol can be estimated directly by gas chromatography if a suitable internal standard is added. A number of procedures have been described, but in most a known amount of lipid is separated by TLC using one of the solvent systems described above, the cholesterol band is located and the compound is eluted from the adsorbent with chloroform. A known amount of cholestane (or octadodecane or desmosterol) is added as internal standard, the solvent is removed and the compounds are taken up in a little diethyl ether for GLC analysis. Columns (50–100 cm long) containing 3–5 per cent SE-30 as lipid phase operated at 200–240°C are generally satisfactory, but more symmetrical peaks may be obtained and a slightly lower column temperature used if the cholesterol is first converted to the acetate or trimethylsilyl ether derivative (see Chapter 4). The amount of cholesterol in the sample is calculated by relating the area of the cholesterol peak on the GLC trace to that of the internal standard. Cholesterol in cholesteryl esters can also be estimated by this method if the latter are first hydrolysed or transesterified. The procedure has been thoroughly evaluated.[539]

Free cholesterol can also be determined by a chemical method.[234]

"Sulphuric acid (10 ml) is added dropwise with stirring to acetic anhydride (60 ml) taking care that the temperature of the mixture does not rise above 10°C. Glacial acetic acid (30 ml) and anhydrous sodium sulphate (0·6 g) are then dissolved in the mixture with care. The reagent (5 ml), which is stable for about two weeks, is added to the lipid in acetic acid solution (0·2 ml) at 25°C and after 20 min the absorbance of the unknown and a series of standards and blanks are determined at 550 or 610 nm. The amount of free cholesterol in the sample is read from a calibration curve prepared from the standards."

Cholesteryl esters must again be hydrolysed before the cholesterol residue can be determined; the procedure works best with pure compounds isolated by chromatography. An excellent alternative procedure for the estimation of cholesterol utilises the reaction of this compound with ferric perchlorate.[604]

C. Alkyl Diglycerides and Neutral Plasmalogens

1. *Separation and identification*

Alkyl diglycerides (glyceryl ethers) and neutral plasmalogens are generally minor components of animal tissues, but may occasionally assume greater significance. Methods of analysis of these compounds have recently been reviewed.[337, 559] In addition, cholesteryl ethers[144] and plasmalogens[153] have been detected in trace amounts in bovine heart muscle but it is not known whether they occur in other tissues or other species.

Adsorption TLC on silica gel layers can be used to separate simple alkyl and alkenyl lipids; neutral plasmalogens tend to migrate ahead of alkyl diglycerides which in turn migrate just in front of triglycerides as illustrated in Fig. 6.5. Complete resolution of each is not always easy but a double development in a single direction with toluene, toluene–methanol (199:1, v/v)[428] or hexane–diethyl ether (95:5, v/v)[465] as solvent system frequently gives satisfactory results. A TLC procedure has been described for separating trace amounts of alkoxy-glycerides from large quantities of triglycerides,[465] but column chromatographic procedures do not in general possess sufficient resolving power.

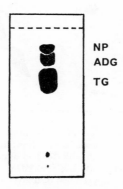

NP
ADG
TG

F<small>IG</small>. 6.5. Thin-layer separation of triglycerides, alkyl diglycerides and neutral plasmalogens on silica gel G layers. Solvent system; hexane–diethyl ether (95:5, v/v; double development in same direction). NP, neutral plasmalogens; ADG, alkyldiglycerides; TG, triglycerides.

Neutral plasmalogens may be detected by spraying the TLC plates with 2,4-dinitrophenylhydrazine (0·4 per cent) in 2 N hydrochloric acid; aldehydes are released which show up as yellow-orange spots on warming the plates. Alternatively a two-dimensional reaction TLC technique can be used; the sample is applied in the bottom left-hand corner and the plate run in the first direction with hexane–diethyl ether (95:5, v/v) as solvent system, the plate is exposed to hydrochloric acid fumes to liberate the aldehydes and then turned anticlockwise and run in the second direction with hexane–diethyl ether (80:20, v/v) as developing solvent. The reaction products are easily distinguished from the unchanged compounds after detection with suitable reagents (see Fig. 6.6)[466] and all can be recovered from the plate for further analysis (although there are doubts as to whether the recovery of aldehydes is quantitative).

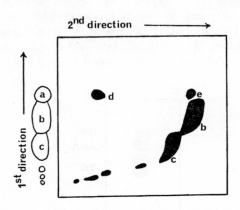

FIG. 6.6. Two-dimensional reaction TLC of neutral plasmalogens and related lipids on silica gel G. Solvent system, 1st direction: hexane–diethyl ether (95:5, v/v). Plate then exposed to hydrochloric acid fumes. Solvent system, 2nd direction: hexane–diethyl ether (80:20, v/v).[466] a, neutral plasmalogens; b, alkyldiglycerides; c, triglycerides; d, diglycerides; e, aldehydes.

Spectroscopic aids to identification are also useful with plasmalogens; the ether-linked double bond exhibits a characteristic band in the IR spectrum at $6\cdot1$ μm[464, 467, 499] and the olefinic protons adjacent to the ether bond produce a doublet centred at $4\cdot11\tau$ in the NMR spectrum.[464, 499]

Alkyl diglycerides are less easily recognised and the ether linkage is not easily disrupted. They must be identified by their chromatographic behaviour on TLC adsorbents relative to an authentic standard or by the chromatographic behaviour of the hydrolysis products—free fatty acids and 1-alkoxy glycerols. The latter compounds migrate close to monoglycerides on TLC adsorbents but cannot be hydrolysed further and react with periodate-Schiff reagent (see Chapter 7).

The ether bond in alkyl diglycerides produces a sharp characteristic band at 9 μm in their IR spectrum.[43, 77] In the NMR spectra of the compounds, the protons on the carbon atom adjacent to the ether group give rise to a characteristic signal in the form of a triplet centred on 6·6τ.[477]

Alkyl diglycerides and plasmalogens may be estimated by charring followed by photodensitometry or better by gas chromatography of the methyl esters of the fatty acid components of the compounds with a suitable internal standard. Basic transesterification procedures should be used to avoid disruption of the vinyl ether bond which would result in contamination of the methyl esters by free aldehydes or dimethyl acetals. The methyl esters should be separated from the other hydrolysis products, which can also be recovered for analysis by the procedures described below, by preparative TLC on silica gel G layers with hexane–diethyl ether (95:5, v/v) as solvent system; methyl esters migrate close to the solvent front and the other products remain near the origin.

As cautioned earlier (Chapter 2), lipid samples containing plasmalogens should not be stored for long periods in solvents containing acetone, methanol or glacial acetic acid as some rearrangement may occur.

2. *Isolation and identification of aldehydes from plasmalogens*

The total amount of plasmalogenic material in a lipid sample can be determined by preparing the *p*-nitrophenylhydrazone derivative of the aldehydes (see Chapter 4 for details of preparation) and estimating these spectrophotometrically at 395 nm relative to a suitable blank. The amount present is obtained from a standard calibration curve prepared at the same time.[135, 421]

If it is intended that individual isomers of the fatty aldehydes be determined, they must first be liberated quantitatively from the neutral plasmalogens. Anderson *et al.*[22] have critically examined many of the acidic hydrolysis procedures that have been described and recommend the following:

"The plasmalogens (0·2–2 mg) in diethyl ether (1·5 ml) are shaken vigorously for 2 min with concentrated hydrochloric acid (1 ml).

The ether layer is removed and the aqueous phase is extracted once more with ether and once with hexane. The combined extracts are washed twice with distilled water before the solvent is removed in a stream of nitrogen. The free aldehydes are obtained by preparative TLC on silica gel G layers with hexane–diethyl ether (90:10, v/v) as solvent system. Aldehydes are found just below the solvent front and can be recovered from the adsorbent for further analysis by elution with diethyl ether. 2,3-Diacyl-*sn*-glycerols, which are the other product of the reaction, are found much further down the plate."

The hydrolysis procedure can be carried out in the presence of silica gel if necessary. It is now known that, although complete hydrolysis of the vinyl ether bond occurs with this method, only 80 per cent recovery of aldehydes can be attained although they are probably representative in composition of those originally present in the natural compound.[592] Aldehydes can also be liberated from plasmalogens with 5 mM aqueous mercuric chloride solution,[391] possibly a milder reagent, although the effect of the reaction on the aldehydes liberated has not been as rigorously tested as with the acid hydrolysis procedure.

Gray[159] and Mahadevan[329] have reviewed GLC procedures for the analysis of aldehydes and their derivatives. Aldehydes can be analysed by GLC without prior conversion to other derivatives on similar columns to those used for methyl ester analysis and they can be identified and estimated by analogous procedures. Standard aldehyde mixtures are available commercially or they can be synthesised from the corresponding fatty acids by a number of methods;[329] for example, preparation from either fatty acids or fatty alcohols is possible.[136] Because of the tendency of free aldehydes to polymerise on standing especially in the presence of traces of alkali, however, it is more usual to convert them to more stable derivatives. None the less, aldehydes have recently been reported to be stable for long periods if stored at $-20°C$ in solution in carbon disulphide[589] or other inert solvents such as pentane or diethyl ether. Of the derivatives, acetals are the most popular (see Chapter 4 for details of preparation) and cyclic acetals, such as those of 1,3-propanediol, have been widely used as they are stable under a wide range of conditions. On the other hand, dimethyl acetals are more easily prepared but may decompose to form alk-1-enyl methyl ethers if aluminium columns or stationary phases containing phosphoric acid are used in the gas chromatograph.[332, 516] If glass columns are used with packing materials that contain catalyst-free stationary phases, artefact formation is negligible and dimethyl acetal derivatives are particularly recommended for the analysis of aldehydes by Gray.[159] They are formed readily both from free aldehydes and from plasmalogens (methyl esters are also formed from the latter but can be eliminated by alkaline hydrolysis) and can be separated according to chain length and number of double bonds under similar GLC conditions as are used with the analogous methyl esters. They can be identified by their retention times relative to authentic standards or by using equivalent chain length (ECL) values as illustrated earlier (Chapter 5) for methyl esters. In addition, individual components can be isolated and characterised by similar procedures to those used to determine the structures of fatty acids.

Dimethylhydrazone derivatives of aldehydes are also easily prepared and are eminently suited to GLC analysis,[247] but they have not been widely used for the purpose.

The IR spectra of free aldehydes are similar to those of the related esters except that the characteristic frequency of the carbonyl function is at $5 \cdot 9$ μm with an additional band at $3 \cdot 7$ μm.[331] A triplet at $0 \cdot 3$ τ is characteristic of the proton on the carbonyl group in NMR spectra of aldehydes.[329] Mass spectra of many aldehydes have also been recorded; though the parent molecular ion is rarely seen, there are characteristic fragments at $m/e =$ M-18, M-28 and M-44.[85] More distinctive spectra, that are more easily interpreted, are obtained from the alcohols prepared from the aldehydes by lithium aluminium hydride reduction (W. R. Morrison, personal communication). The melting points of most aldehydes are too low to be of assistance in their identification, but those of the 2,4-dinitrophenylhydrazone derivatives of fatty aldehydes are often sufficiently distinct.[331]

Free aldehydes should not be stored in contact with other lipids, especially those containing ethanolamine, which catalyses a condensation reaction in which 2,3-dialkylacroleins are formed.[470]

3. *Isolation and identification of* 1-*alkyl glycerols*

1-Alkyl glycerols are released from alkyl diglycerides by saponification or transesterification but better yields are obtained if hydrogenolysis with lithium aluminium hydride is used.[537, 596] The procedure also removes the phosphorus group from 1-alkyl phosphoglycerides and, in addition, alk-1'-enyl-1-glycerols are liberated from plasmalogens. The following method can be recommended.[596]

"Lithium aluminium hydride (15 mg) in diethyl ether (3 ml) is added to the lipid (5 mg) in diethyl ether (0·5 ml) in a test-tube and the solution is refluxed for 30 min. On cooling, water (3 ml) is cautiously added followed by 4 per cent acetic acid (3 ml) and diethyl ether (3 ml) and the mixture is thoroughly shaken. The ether layer is removed by Pasteur pipette (after centrifugation to break any emulsions if necessary) and the aqueous layer is extracted twice more with diethyl ether (6 ml portions). The solvent is removed from the combined extracts in a stream of nitrogen (the last traces of water are removed *in vacuo*) and the samples are dissolved in a little chloroform and applied to a silica gel G TLC plate which is then developed in diethyl ether—30 per cent aqueous ammonia (400:1, v/v). The products are identified by their R_f values relative to standards and eluted from the adsorbents with several volumes of diethyl ether."

The alkoxy- and alkenyloxy-glycerols can be estimated on the TLC plate by charring and photodensitometry[596] or they can be recovered from the adsorbent and analysed separately by appropriate procedures. Alk-1'-enyl-1-glycerols are not normally analysed in this form but are converted to aldehydes as described above. None the less, the diacetyl derivatives of these compounds have been prepared and analysed by gas chromatography on SE-30 columns.[17] The fatty acid components of the sample are reduced to fatty alcohols during the reaction but can be recovered for analysis as described below.

Alkyl-1-glycerols must be converted to less polar volatile derivatives such as isopropylidene compounds,[187] trimethylsilyl ethers,[594] trifluoroacetates[594] or acetates[17] for GLC analysis (see Chapter 4 for details of preparation). All give satisfactory results, but isopropylidene derivatives appear to be generally favoured. They may be separated both according to chain length and to the number of double bonds in the alkyl chain on GLC columns packed with similar polyester liquid phases, e.g. EGS, EGSS-X or EGSS-Y, as are used to separate methyl esters except that slightly higher column temperatures are necessary. Suitable standard compounds are available commercially or may be synthesised with fatty acids or alcohols as starting materials by well-tried procedures of which that of Baumann and Mangold[42] is probably the most versatile. Individual isomers of natural alkyl-1-glycerols or their derivatives can be isolated by procedures related to those described for fatty acids[594] and the positions of double bonds can be determined by permanganate–periodate oxidation[187] or by ozonolysis[418] with GLC identification of the fragments.

Of the many other methods of analysing alkyl-1-glycerols that have been described, one other merits further consideration; the preparation under basic conditions of thionocarbonate derivatives which are estimated by their absorbance at 235 nm and which are also suited to GLC analysis[417] (it is also suggested that alk-1'-enyl-1-glycerols may be analysed in this form).

IR and NMR spectra of alkyl-1-glycerols or their derivatives are similar to those of the parent alkyl diglycerides except that the free hydroxyl groups or specific functional groups in the derivatives introduce additional features.[595] Individual isomers can also be identified by mass spectrometry[180] and by their critical solution temperatures.[469]

D. Wax Esters and Related Compounds

Natural waxes may contain a wide variety of simple lipid components including hydrocarbons, wax esters, fatty alcohols and ketones, sterols and sterol esters as well as the more common simple lipids such as triglycerides. Thin-layer chromatography on layers of silica gel has proved to be the best method of separating these compounds; silicic acid columns lack resolution, but may on occasion be suited to specific natural mixtures. Holloway and Challen[211] have systematically studied TLC methods for separating wax constituents and have developed several valuable solvent systems although no single one will adequately resolve all compounds.

Hydrocarbons tend to elute near the solvent front with even the least polar of solvent systems, but normal paraffins are eluted just ahead of squalene if carbon tetrachloride[211] or hexane–benzene (9:1, v/v)[174] are the developing solvents. Individual hydrocarbon isomers can be separated by gas chromatography; for example on a capillary column linked to a mass spectrometer, 93 different compounds from wool wax have been separated and identified.[352] Homologous series of hydrocarbons differing in degree of unsaturation have been separated by silver nitrate TLC.[511]

Wax esters are only slightly more polar than cholesteryl esters from which they are not easily separated. Double development in carbon tetrachloride[211] or hexane–diethyl ether (98:2 v/v) on layers of silica gel G may be successful, however, although there is frequently some band spread because of partial separation into molecular species.

Primary fatty alcohols migrate just above free cholesterol with the solvent systems described in Fig. 6.1 and secondary alcohols migrate just ahead of the primary compounds. Morris *et al.*[363] have described the TLC behaviour of the complete series of hydroxy-octadecanes; only the isomers with hydroxyl groups in positions 1, 2 or 3 are separable from the remainder.

Of the compounds generally classed as wax components, wax esters, which consist of long-chain fatty acids esterified to fatty primary alcohols of analogous structure, and their constituents have been most studied as they are major components of the lipids of many marine animals. The fatty acids and alcohols are liberated by alkaline hydrolysis and may be separated from each other by solvent extraction of the alkaline solution (see Chapter 4) or by TLC. Subsequently, individual isomers of the fatty alcohols, usually in the form of less polar derivatives to minimise tailing, can be separated by chain length and by degree of unsaturation and estimated by GLC on similar polyester liquid phases as are used with methyl esters. Acetate derivatives are generally preferred as the resolution of individual isomers on polyester columns is somewhat better than that obtained with trifluoroacetate or trimethylsilyl ether derivatives,[241] although these are eluted earlier than acetates and are sometimes also used. It should be noted that the acetate derivatives of polyunsaturated fatty alcohols do not necessarily elute in the same order as the corresponding methyl esters on polyester columns.[241]

Methyl esters of fatty acids are occasionally found as constituents of tissues, but as they may be formed as artefacts during extraction of tissues with solvents containing methanol, the natural origin of such compounds must be established (see Chapter 2) when their presence is detected.

CHAPTER 7

The Analysis of Complex Lipids

COMPLEX lipids from natural sources may contain such a multiplicity of components that complete separation, identification and estimation of each one is an extremely difficult task. Indeed, it is not possible to describe a definitive procedure that can be used with confidence in all circumstances. Commonly, the total lipid extract from a tissue is subjected to a preliminary separation by column chromatography on silicic acid and the complex lipids separated from simple lipids. Single and two-dimensional TLC systems have been developed that permit separation of most complex lipids on an analytical or semi-preparative scale, but when larger amounts of individual components are required for more detailed analysis, column chromatography on DEAE cellulose, for example, must be used to obtain simpler fractions that can be further subdivided by column or one-dimensional TLC techniques. When only one compound or group of compounds is required, short cuts may be available. Unequivocal identification of complex lipids requires that the various hydrolysis products from each compound be identified and estimated. Precautions should always be taken to minimise the effects of autoxidation (see Chapter 3).

A. Preliminary Separation and Preparation of Lipid Samples

1. Simple group separations

The following procedures apply to samples of complex lipids that do not contain gangliosides, i.e. they are obtained by the conventional extraction-washing procedures described in Chapter 2 (and also below).

Complex lipids can be separated from simple lipids by column chromatography on silicic acid or acid-washed Florisil (30 mg lipid/g adsorbent) as described in Chapter 6. Briefly, chloroform or diethyl ether (10 column volumes) will elute the simple lipids and methanol (10 column volumes) the complex lipids. For simple fractionations of this nature, short wide columns are used so that rapid flow rates are achieved. If necessary further simplification of natural mixtures is possible, as glycolipids can be recovered as a class by elution with acetone after the simple lipids have been eluted from the column. In particular, mono- and digalactosyl diglycerides and plant sulpholipid may be separated from the phospholipids of plant tissues[565] and glycosyl diglycerides can be isolated from bacterial lipids[480] by elution with this solvent (10 column volumes). Acetone or acetone–methanol (9:1, v/v; 40 column volumes) will also elute ceramide monohexosides, ceramide polyhexosides and sulphatides from lipid extracts of animal tissues[445] and tetrahydrofuran–methylal–methanol–water (10:6:4:1 by volume) has been used with equal effect.[160] Phospholipids, virtually free of glycolipids, can then be recovered from the column by elution with methanol.

The principal disadvantage of the procedure is that small amounts of the less polar acidic phospholipids such as cardiolipin and phosphatidic acid, occasionally accompanied by phosphatidyl ethanolamine, may contaminate the acetone fraction. With some batches of adsorbent, greater amounts of these may be eluted with acetone than is acceptable but, when this is seen to occur, it can be minimised by eluting with chloroform–acetone mixtures rather than with acetone alone. In addition, acetone has a strong dehydrating effect on the adsorbent causing the phospholipids to be retained more strongly than otherwise would be the case so that recoveries are sometimes less than ideal.

Individual glycolipids can often be further resolved by eluting with chloroform containing various proportions of acetone while individual phospholipids are recovered subsequently by eluting with chloroform containing methanol and this is discussed in greater detail below.

Simple and complex lipids can also be effectively separated from each other by solvent partition procedures and that described by Galanos and Kapoulas[145] can be especially recommended as a rough preparative method.

"Hexane is equilibrated with 87 per cent ethanol. The hexane layer is contained in two separating funnels (45 ml in each) and up to 10 g of lipid in the ethanolic phase (15 ml) is added to the first funnel. After thorough shaking, the bottom layer is run into the second funnel and the two phases are again mixed after which the bottom layer is run off. A further portion (15 ml) of fresh ethanolic phase is added to the first funnel and the extraction procedure is repeated. Six further portions of fresh ethanolic phase are shaken with the hexane layer in this way. The combined ethanolic layers then contain all the complex lipids and the combined hexane layers most of the simple lipids."

The method is particularly suited to lipid samples that contain a high proportion of simple lipids, for example adipose tissue extracts or milk fats which may contain less than 1 per cent of the total lipid in the form of complex lipids. Emulsion formation can be troublesome on occasion, but the only other drawback of the method is that free fatty acids partition fairly evenly between the two phases. These and traces of other simple lipids can be removed later by column chromatography as described above.

Simple and complex lipids can also be separated by dialysing lipid extracts in hexane solution through a rubber membrane against fresh hexane. Simple lipids, including free fatty acids in this instance, pass through the membrane but complex lipids do not.[546]

2. *Cations associated with complex lipids*

It is not always realised that the solubility and chromatographic properties of acidic phospholipids are dependent to a considerable degree on the cations with which they are associated. In particular, divalent ions, for example those of calcium or magnesium, confer quite different elution characteristics on acidic lipids such as phosphatidic acid, phosphatidyl serine and phosphatidyl inositol on adsorption chromatography from those of the sodium or potassium salts. Single lipid classes may on occasion be split into two or three bands by TLC according to the types of cation associated with the molecules. Before complex lipids are separated into their constituents, it may therefore be advantageous to ensure that all the phospholipid classes are in a single salt form—preferably as the sodium or potassium salts. This can best be achieved by passing the lipid extracts through a chelating resin as described below.[79] Washing the lipid in an appropriate solvent with a solution containing the chosen cation is only partially effective.

"The resin (Chelex 100, 50–100 mesh, Biorad Laboratories) is prepared in the proper ionic form by washing it twice with two volumes of 2 N hydrochloric acid then thoroughly with deionised water. The cycle is repeated three times with 2 N sodium hydroxide solution (or potassium hydroxide depending on which cation is required) and then the resin is washed with deionised water until its pH is approximately 12. The pH is adjusted to 8·0 by washing with 2 N acetic acid when the solvent is changed to chloroform–methanol–water (5:4:1 by volume) by thorough washing. The lipids (up to 10 g) are added to a column (4 × 50 cm) of the resin in this solvent which is also used for elution (3 column volumes). Complete conversion to the required salt form occurs."

Although this procedure may not be necessary with all lipid samples and is not practical when large numbers of samples have to be analysed routinely, it should be accepted as general practice when highly precise analysis is intended.

B. Column Chromatography

Complex lipids eluted from columns can be determined by the general procedures described in Chapter 3. In addition, several more specific methods are available, for example determination of phosphorus in glycerophosphatides or of nitrogen in glycosphingolipids, and these methods are discussed in detail below.

1. *Silicic acid*

Chromatography on columns of silicic acid is a useful preparative procedure for subdividing large amounts of complex lipids into simpler fractions. Pure single compounds are not easily obtained from natural

G

mixtures by this technique, however, and it is customary to use the method in conjunction with preparative TLC or with DEAE cellulose chromatography for complete separations. The preparation of suitable columns has been described in Chapter 3 and the application of the method to the analysis of complex lipids has been reviewed elsewhere.[446, 529]

The principal advantages of the method lie in the ease of preparation of the column and in the comparatively large amount of lipid that can be separated (30 mg lipid per g of adsorbent). On the debit side, the properties of the adsorbent may change with slight changes in its physical state, e.g. in the size of the particles or in the degree of hydration (there are signs that manufacturers of adsorbents are close to the solution to the problem). As a result, the acidic lipids such as phosphatidyl serine or phosphatidyl inositol may change their elution characteristics with respect to non-acidic lipids. With chloroform containing various proportions of methanol, acidic lipids are eluted in the order:

cardiolipin
  phosphatidic acid
    cerebroside sulphate or plant sulpholipid
     phosphatidyl glycerol
      phosphatidyl serine
       phosphatidyl inositol
        di- and tri-phosphoinositides

and non-acidic complex lipids in the order:

ceramide
  ceramide monohexoside and glycosyl diglycerides
    phosphatidyl ethanolamine
     phosphatidyl choline, lysophosphatidyl ethanolamine, ceramide dihexoside
      sphingomyelin
       lysophosphatidyl choline

The elution characteristics of the acidic lipids are also influenced by the cations with which they are associated. Both groups of lipids are always eluted in a constant order, but the regions where the two groups overlap may vary somewhat and it may not always be possible to obtain clear-cut fractions. In consequence, it is usually necessary to determine the elution characteristics of a fresh batch of adsorbent before proceeding. As a rough guide, phosphatidyl ethanolamine and phosphatidyl serine tend to elute together while phosphatidyl inositol elutes just ahead of phosphatidyl choline.

More useful separations can be obtained if chloroform–acetone mixtures are used to elute glycolipids before phospholipids are recovered by elution with chloroform–methanol mixtures. With plant lipids,[565] for example, monogalactosyl diglycerides are recovered free of most other complex lipids by elution with chloroform–acetone (1:1 v/v; 5 column volumes); acetone alone (10 column volumes) will then elute digalactosyl diglycerides and plant sulpholipid together with some of the less polar phospholipids such as cardiolipin or phosphatidic acid although the latter are not often major components of plant lipids. Chloroform–acetone mixtures are occasionally used with animal lipids[449] but are generally less useful as glycolipids tend to be minor components (with the important exception of brain lipids), but acetone alone (40 column volumes) can be usefully employed to elute cerebroside, cerebroside sulphate and ceramide dihexoside as discussed above.

Phospholipids can then be recovered from the column by elution with chloroform containing increasing amounts of methanol. In a typical separation, phosphatidic acid and cardiolipin are eluted first with the solvents in the ratio 95:5, v/v (10 column volumes) followed by phosphatidyl ethanolamine and phosphatidyl serine (80:20, v/v; 20 column volumes), phosphatidyl choline and phosphatidyl inositol (50:50, v/v; 20 column volumes) and finally sphingomyelin and lysophosphatidyl choline are eluted with methanol alone. If care is taken and very slow flow rates, low lipid loads and gradient elution techniques are used, much finer separations are possible,[369, 372] but these are usually attainable with less effort by other methods. The nature of the fractions obtained can be monitored by means of micro-TLC plates (see Chapter 3).

Acid-washed silicic acid (prepared by washing the adsorbent with 6 N hydrochloric acid followed by water and activation at 120°C) can sometimes be used to obtain separations that are not possible with the untreated adsorbent. For example, phosphatidyl serine is eluted before rather than with phosphatidyl ethanolamine on acid-washed silicic acid.[446]

Silicic acid–silicate columns also are a useful alternative to the untreated adsorbent. A column prepared in the usual manner is washed with chloroform–methanol (4:1, v/v; 1 column volume) containing 1 per cent ammonia. Phosphatidyl ethanolamine is eluted with chloroform–methanol (4:1, v/v; 8 column volumes) and phosphatidyl serine with methanol (3 column volumes) from such a column.[448] Phosphatidyl choline and sphingomyelin are also more easily separated on silicic acid prepared in this way than on the unmodified adsorbent.[444] Table 7.1 contains a summary of some useful procedures.

In short, therefore, silicic acid column chromatography is best considered as a preparative tool for obtaining in quantity fractions enriched in particular lipid components that can then be purified by other methods. It is not a precise analytical procedure. The most useful features are that very little preparatory work is necessary in packing and conditioning columns and that glycolipids and phospholipids can be rapidly separated. Acid-washed Florisil is used in a similar manner to that described above to obtain analogous separations.

2. *Florisil*

Florisil itself is little used for the separation of phosphatides as they are very strongly retained by the adsorbent and are recovered only with difficulty especially when the water content of the adsorbent is low. This phenomenon can be utilised, however, in the isolation of certain glycolipids if very dry solvents (obtained by incorporating 5 per cent 2,3-dimethoxypropane) are used. For example, ceramide is eluted from Florisil with chloroform–methanol–dimethoxypropane (95:5:5 by vol.; 10 column volumes) and cerebroside and cerebroside sulphate with chloroform–methanol–dimethoxypropane (70:30:5 by vol.; 20 column volumes).[442] Glycosyl diglycerides from plant lipids can also be separated from phosphoglycerides by a related procedure.[386]

Such methods should be considered primarily as a means of obtaining concentrates of particular glycolipids rather than as precise analytical

TABLE 7.1. CHROMATOGRAPHY OF COMPLEX LIPIDS ON SILICIC ACID COLUMNS

Fractions	Compounds eluted	Solvents*	Column volumes
Procedure A	Animal lipids on silicic acid		
1	simple lipids	chloroform	10
2	glycolipids with traces of acidic phospholipids	acetone	40
3	phospholipids	methanol	10
Procedure B	Fraction 3 from procedure A on silicic acid		
1	any remaining phosphatidic acid and cardiolipin	chloroform–methanol (95:5 v/v)	10
2	phosphatidyl ethanolamine+phosphatidyl serine	,, (80:20 v/v)	20
3†	phosphatidyl inositol+phosphatidyl choline	,, (50:50 v/v)	20
4	sphingomyelin and lysophosphatidyl choline	methanol	20
Procedure C	Fraction 2 from procedure B on silicic acid-silicate[448]		
1	phosphatidyl ethanolamine	chloroform–methanol (80:20 v/v)	8
2	phosphatidyl serine	methanol	3
Procedure D	Plant lipids on silicic acid[565]		
1	monogalactosyl diglyceride	chloroform–acetone (50:50 v/v)	8
2	digalactosyl diglyceride and sulpholipid	acetone	10
3–6	as procedure B		

* The solvent mixtures described should be taken as an approximate guide only.

† Fractions 3 and 4 of procedure B can be better resolved on occasion with silicic acid-silicate as adsorbent.[444]

techniques as the subsequent recovery of phospholipids is poor. Short wide columns and rapid flow rates can then be used without significant loss of resolution.

3. *Alumina*

Alumina is rarely used as an adsorbent for column chromatography of complex lipids because of its basic nature; hydrolysis of glycerophosphatides occurs and significant amounts of lysophosphatides not originally present in the sample are found. None the less, neutral alumina (Brockman grade IV) is occasionally used as a rapid method of obtaining concentrates of specific lipids. For example, neutral alumina (Merck, A.G., Germany; heated at 110°C for 24 hr then deactivated by adding 10 per cent water) has been used as a rapid means of obtaining phosphatidyl choline[486] and phosphatidyl inositol[317] from lipid samples. Phospholipids are eluted in a quite different order from that described above for silicic acid, i.e.

phosphatidyl choline
  lysophosphatidyl choline and sphingomyelin
    phosphatidyl ethanolamine and lysophosphatidyl ethanolamine
      cardiolipin and phosphatidyl inositol
        phosphatidyl serine

The non-acidic lipids are eluted with chloroform, methanol and water in various proportions, but it is necessary to add ammonium salts to the solvents to recover the acidic lipids.[317]

Batches of alumina may vary much more markedly in their physical properties than those of silicic acid, so solvent systems quoted in the literature can only be taken as an approximate guide. However, smaller volumes of eluting solvents can be used than is required with silicic acid and comparatively rapid flow rates (up to 10 ml/min) are permissible. Although some hydrolysis of lipids inevitably occurs even with deactivated adsorbents, the lipids recovered appear to be identical in structure to the original compounds (i.e. hydrolysis is random) and are suitable for analyses of molecular species.

4. *DEAE and TEAE cellulose chromatography*

DEAE cellulose chromatography is a valuable alternative column procedure to conventional adsorption chromatography for the separation of complex lipids. The principle of the method, the preparation of the adsorbent and packing of the columns are discussed in Chapter 3. The nature of the separations is very different from that of the procedures just described: choline- and ethanolamine-containing lipids are separately eluted from the columns with chloroform–methanol mixtures; weakly acidic lipids such as phosphatidyl serine are eluted with acetic acid and more highly acidic or ionic lipids such as phosphatidic acid or phosphatidyl inositol are eluted with solvents to which inorganic salts or ammonia are added. Table 7.2 lists some useful general elution schemes for animal and plant lipids (bacterial lipids can also be separated efficiently by such procedures).[70] Clear-cut fractions are obtained with virtually no cross-contamination and when more than one compound is present in any given fraction, they are usually separated comparatively easily by subsequent thin-layer chromatography or by other column procedures. Different batches of DEAE cellulose do not vary markedly in their properties so that the elution schemes described in Table 7.2 are trustworthy. As a result, bulk fractions can be taken with almost no danger of fraction overlap even when the volumes of solvent used are increased.

DEAE cellulose acetate procedures can also be used as a comparatively large-scale procedure for isolating acidic lipids. For example, 0·5–0·6 g of phosphatidyl serine can be isolated from extracts of brain tissue (50 g) on a 2×14 cm column of DEAE cellulose in a very short time.[458]

With plant lipids, a preliminary fractionation of glyco- and phospholipids on silicic acid columns as described above is unnecessary if DEAE cellulose chromatography is used as the glycolipids are eluted in separate distinct fractions. Indeed, no preliminary separation of simple and complex lipids is necessary as simple lipids, with the exception of free fatty acids, can be eluted with chloroform (10 column volumes) before the elution sequences in Table 7.2 are commenced (with plant lipids, simple lipids are eluted in the first 10 column volumes of chloroform and monogalactosyl-diglycerides in the second 10 column

TABLE 7.2. ELUTION SCHEMES FOR DEAE CELLULOSE COLUMN CHROMATOGRAPHY

Fraction	Lipids eluted	Solvents	Column volumes
A	*Animal lipids*[446]		
1	phosphatidyl choline lysophosphatidyl choline sphingomyelin cerebroside	chloroform–methanol (9:1, v/v)	10
2	phosphatidyl ethanolamine ceramide di- and poly-hexosides lysophosphatidyl ethanolamine	chloroform–methanol (1:1, v/v)	10
3	—	methanol	10
4	phosphatidyl serine	glacial acetic acid	10
5	—	methanol	4
6	cardiolipin phosphatidic acid phosphatidyl glycerol phosphatidyl inositol cerebroside sulphate	chloroform–methanol (4:1, v/v) made 0·05 M with respect to ammonium acetate to which is added 20 ml of 28 per cent ammonia/litre	10
7	—	methanol	10

TABLE 7.2 (*continued*)

Fraction	Lipids eluted	Solvents	Column volumes
B	*Plant lipids[375]*		
1	monogalactosyl diglyceride	chloroform	20
2	{ phosphatidyl choline cerebroside sterol glycoside	chloroform-methanol (95:5, v/v)	10
3	digalactosyl diglyceride	chloroform-methanol (90:10, v/v)	10
4	phosphatidyl ethanolamine	chloroform-methanol (60:40, v/v)	10
5	{ cardiolipin phosphatidic acid phosphatidyl glycerol phosphatidyl inositol plant sulpholipid	solvent as for fraction 6 above	10
6	—	methanol	10

volumes). Free fatty acids are eluted with the weakly acidic lipids (fraction 4 in scheme A and fraction 5 in scheme B), but if they are major components of the sample, a preliminary separation of the simple lipids on a silicic acid column is advantageous as otherwise the retention power of the DEAE cellulose adsorbent for the other acidic lipids is diminished.

Care should be taken not to change back from polar to non-polar solvents too quickly as the column packing may be disturbed and the quality of the separations reduced. For this reason, methanol washes are inserted as fractions 5 and 7 in elution scheme A (Table 7.2). If the column is not allowed to run dry, it can be regenerated for further use after the final methanol wash by elution with glacial acetic acid (3 column volumes), methanol (3 column volumes), chloroform–methanol (1:1, v/v; 3 column volumes) and chloroform (4 column volumes). With comparatively simple lipid mixtures, the solvent is run rapidly (approximately 10 ml/min) through the column, but with more difficult samples a slower flow rate (3 ml/min) may be preferable. Salts in column fractions can be removed by Sephadex chromatography or by a "Folch" wash (see Chapter 2).

DEAE cellulose in the borate form[446] has somewhat different properties; it is prepared by washing the DEAE cellulose column prepared as described earlier (Chapter 3) with a saturated solution of sodium tetraborate in methanol (3 column volumes) after which excess borate is removed by washing with methanol then chloroform–methanol mixtures. The chief virtue of such columns is that phosphatidyl ethanolamine is more strongly retained so that ceramide polyhexosides are obtained as a separate class free of glycerophosphatides by elution with chloroform–methanol (2:1 v/v; 10 column volumes). After washing the column with methanol, pure phosphatidyl ethanolamine is recovered by elution with chloroform–methanol–acetic acid (63:33:1 by vol.; 8 column volumes). Such columns are accordingly a useful supplement to conventional DEAE cellulose chromatography as they permit the subdivision of the second fraction (scheme A, Table 7.2) into simpler components. Columns must be regenerated with sodium tetraborate in methanol before they are reused.

TEAE cellulose in the hydroxyl form is also a useful alternative to DEAE cellulose as it has a much higher capacity for acidic lipids and as ceramide polyhexosides and phosphatidyl ethanolamine are eluted in quite different fractions. Columns are prepared as follows:[446]

"The TEAE cellulose adsorbent is prepared and packed into columns in a slurry of acetic acid as described earlier (Chapter 3) for DEAE cellulose. It is then washed with methanol (3 column volumes) and converted to the hydroxyl form by washing with 0·1 N potassium hydroxide in methanol (4 column volumes). After further washes with methanol (8 column volumes), chloroform–methanol (1:1 v/v; 4 column volumes) and chloroform (4 column volumes), the column is ready for use."

Approximately 10–30 mg of lipid per g of adsorbent can be separated on this column if a solvent flow rate of 3 ml/min is not exceeded. The recommended elution sequence is outlined in Table 7.3. In this instance, fraction 2 contains only ceramide di- and polyhexosides and phosphatidyl ethanolamine is eluted cleanly in a later fraction. When the former group of compounds are of particular interest to the analyst, therefore, TEAE cellulose chromatography can be recommended.[544, 545]

Column chromatography on DEAE cellulose in the acetate form or TEAE cellulose in the hydroxyl form are probably the most useful preparative methods for simplifying complex lipid mixtures. The only disadvantages are the rather lengthy conditioning necessary before columns can be used, some oxidation of phosphatidyl ethanolamine may occur and any phosphatidyl serine plasmalogens present are destroyed by the acidic conditions necessary to elute them from the columns. The excellent review articles of Rouser *et al.*[446, 447] are mandatory reading before the techniques are used.

C. Thin-layer Chromatography

Similar separations to those about to be described can be obtained on silicic acid-impregnated paper,[134, 263] but for the reasons given earlier (Chapter 3), thin-layer chromatography is generally preferred. TLC of complex lipids has been reviewed elsewhere.[437, 489]

TABLE 7.3. ELUTION SCHEME FOR TEAE CELLULOSE CHROMATOGRAPHY

Fraction	Lipids eluted	Solvents	Column volumes
1	As in fraction A-1 (Table 7.2)	chloroform–methanol (2:1, v/v)	8
2	ceramide polyhexosides		8
3	—	methanol	8
4	phosphatidyl ethanolamine phosphatidyl N-methylethanolamine phosphatidyl NN-dimethyl-ethanolamine	chloroform–methanol–glacial acetic acid (63:3:1, by vol.)	6
5	phosphatidyl serine	glacial acetic acid	6
6	—	methanol	3
7	As in fraction A-6 (Table 7.2)		6

1. *Single-dimensional TLC systems*

Non-acidic phospholipids may be separated on plates coated with thin layers of silica gel G developed in a solvent system of chloroform–methanol–water (25:10:1 by vol.). Lipids migrate in a similar order to the elution sequence outlined above for silicic acid columns and a typical separation is illustrated in Fig. 7.1 (plate A). Satisfactory separations are achieved as long as acidic lipids are not present in significant

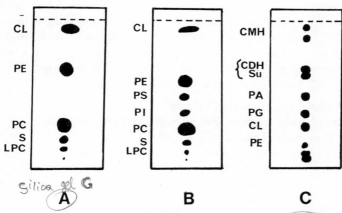

Fig. 7.1. TLC separations of complex lipids from animal tissues. Plate A, Silica gel G. Plates B and C, Silica gel H. Solvent systems: plate A, chloroform–methanol–water (25:10:1 by vol.); plate B, chloroform–methanol–acetic acid–water (25:15:4:2 by vol.);[491] plate C, 1st development–pyridine–hexane (3:1, v/v); 2nd development—chloroform–methanol–pyridine–2 M ammonia (35:12:65:1 by vol.).[490] Abbreviations: CL, cardiolipin; PE, phosphatidyl ethanolamine; PC, phosphatidyl choline; S, sphingomyelin; LPC, lysophosphatidyl choline; PS, phosphatidyl serine; PI, phosphatidyl inositol; CMH, ceramide monohexosides; CDH, ceramide dihexosides; Su, sulphatide; PA, phosphatidic acid; PG, phosphatidyl glycerol.

amounts, otherwise phosphatidyl serine will contaminate the phosphatidyl ethanolamine while phosphatidyl inositol will contaminate the phosphatidyl choline. It is a particularly useful TLC system for isolating individual components from fractions 1 and occasionally 2 of the general DEAE cellulose elution scheme for animal lipids outlined in Table 7.2 or for isolating the products of phospholipase A hydrolysis of phospholipids (see Chapter 9). Up to 10 mg of phospholipids can be applied to a 20 × 20 cm plate coated with a layer 0·5 mm thick of silica gel G. Variations in the properties of the adsorbent are countered by raising or lowering the amount of water in the solvent system. When they are present in the sample, simple lipids migrate with the solvent front.

There is no simple TLC system that can be used to separate all known complex lipids in a single dimension, but Skipski *et al.*[491] have developed a useful procedure for the more common acidic and neutral glycerophosphatides. Plates are coated with a layer of silica gel H (without calcium sulphate as binder) made in a slurry of 1 mM sodium carbonate solution to render it basic. After the plates are activated, lipids are applied in the usual way and the plates are developed in a solvent system consisting of chloroform–methanol–acetic acid–water (25:15:4:2 by vol.). Phospholipids migrate in the order:

cardiolipin and phosphatidic acid
phosphatidyl glycerol and phosphatidyl ethanolamine
phosphatidyl serine
phosphatidyl inositol
phosphatidyl choline
sphingomyelin
lysophosphatidyl choline

When the acidic lipids phosphatidyl serine and phosphatidyl inositol are not completely separated from the non-acidic lipids because of slight differences in the properties of the adsorbent, small changes in the proportions of acetic acid and water in the solvent mixture will generally rectify the situation. Figure 7.1 (plate B) illustrates the nature of the separations that can be obtained with a natural mixture, pig liver lipids, which contains these components. Simple lipids run with the solvent front, but better results are obtained if they are first moved to the top of the plate by a prewash with acetone–hexane (1:3, v/v)[489] and the plate dried at room temperature in a vacuum oven before the phospholipids are separated as above. It has been claimed[271a] that the addition of 0·4 per cent ammonium sulphate to the silica gel H layer and a slight change in the proportions of the solvents (to 50:25:8:1 by vol.) improves the reproducibility of the system.

The principal disadvantages of the procedure are that phosphatidic acid and cardiolipin and phosphatidyl glycerol and phosphatidyl ethanolamine are not separated and in addition glycosphingolipids may contaminate the phosphatidyl ethanolamine. These difficulties can be partly overcome by a two-step single-dimensional TLC system.[490] Lipids are applied to a silica gel H plate prepared as above which is first developed in pyridine–hexane (3:1 v/v) to the top when it is removed from the solvent tank and dried in a vacuum oven at room temperature. It is then redeveloped in the same direction with chloroform–methanol–pyridine–2 M ammonia (35:12:65:1 by vol.) as solvent system until the solvent is about 2 cm below the previous solvent front. With this system, lipids migrate in the order:

cerebrosides
   ceramide dihexosides and sulphatides
      phosphatidic acid
         phosphatidyl glycerol
           cardiolipin
              phosphatidyl ethanolamine

as illustrated in Fig. 7.1 (plate C). Fractions 2 and 6 of the DEAE cellulose elution scheme (Table 7.2A) can therefore be separated into individual components with this system. More useful separations of complex glycosphingolipids may be attained by TLC if these compounds are first separated from glycerophosphatides by an appropriate procedure. Such methods have been mentioned briefly and are discussed in greater detail below.

Fractions enriched in acidic lipids such as phosphatidyl serine or phosphatidyl inositol obtained from DEAE cellulose columns are best purified on layers of silica gel containing magnesium silicate (10 per cent, w/w) as binder;[443] chloroform–methanol–7 N ammonia (60:35:4, by vol.) is a suitable developing solvent in this instance. Polyphosphoinositides can be resolved on silica gel H layers to which 2·5 per cent potassium oxalate has been added (to sequester any calcium) with chloroform–methanol–4 N ammonia (9:7:2 by vol.) as solvent system.[156]

Single-dimensional TLC systems are used less with plant lipids as glycolipids tend to overlap with phospholipids when chloroform–methanol solvent systems are used. None the less, solvents containing acetone may on occasion offer useful separations. For example, acetone–acetic acid–water (100:2:1 by vol.) has been used to separate mono- and digalactosyl diglycerides from other complex lipids on layers of silica gel G (as illustrated in Fig. 7.2, plate A).[147] More comprehensive separations have been described by Nichols (Fig. 7.2, plate B)[373] on silica gel G layers with diisobutyl ketone–acetic acid–water (40:25:3.7 by vol.) as developing solvent and further useful systems have been described by others.[103, 398]

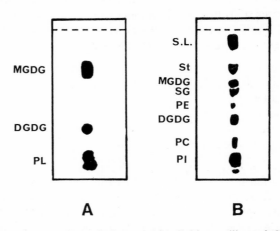

Fig. 7.2. TLC separations of plant complex lipids on silica gel G. Solvent systems: plate A, acetone–acetic acid–water (100:2:1 by vol.);[147] plate B, diisobutyl ketone–acetic acid–water (40:25:3·7 by vol.).[373] Abbreviations: MGDG, monogalactosyl diglycerides; DGDG, digalactosyl diglycerides; PL, phospholipids; SL, simple lipids; St, sterols; SG, sterol glycosides; PE, phosphatidyl ethanolamine; PC, phosphatidyl choline; PI, phosphatidyl inositol.

Methods of recovery of complex lipids from TLC adsorbents have been discussed earlier (Chapter 3 and 6) and methods of detecting, identifying and estimating complex lipids are described in detail below. If single-dimensional TLC systems are to be used for routine separations of complex lipids, the identity and purity of components should be periodically checked by two-dimensional TLC to confirm that single spots do not conceal more than one compound.

2. Two-dimensional TLC systems

Many more complex lipid components can be separated on a single TLC plate if two-dimensional systems are used (see Chapter 3). Very many such systems have been devised and Rouser et al.[446] in particular have described a variety of solvent combinations suited to the analysis of animal lipids. The most successful separations are achieved when contrasting solvents are used for development in each direction; for example, an acidic solvent mixture in the first direction may be followed by development with a neutral or a basic solvent mixture in the second direction, or the second system may contain acetone to retard the migration of phospholipids relative to glycolipids with which they might otherwise overlap. Four solvent systems are favoured by Rouser:

(a) chloroform–methanol–water (65:25:4 by vol.).
(b) n-butanol–acetic acid–water (60:20:20 by vol.).
(c) chloroform–methanol–28 per cent aqueous ammonia (65:35:5 by vol.).
(d) chloroform–acetone–methanol–acetic acid–water (10:4:2:2:1 by vol.).

In Fig. 7.3, plate A illustrates the type of separation possible with solvent a in the first direction and solvent b in the second; plate B has solvent c in the first direction and solvent d in the second. Many other permutations of these solvent systems are possible and more than thirty different complex lipids have been resolved in this way. Silica gel H is usually the favoured adsorbent, but Rouser et al.[443] prefer silica gel containing magnesium silicate (10 per cent w/w) as binder as this apparently gives more compact spots particularly with acidic lipids. Gray[161] also has described a useful TLC system in which solvent mixture a above is used in the first direction and tetrahydrofuran–methylal–methanol–2 M aqueous ammonia (10:5:5:1 by vol.) is used in the second direction (plate C, Fig. 7.3). Unfortunately solvents such as butanol are not easy to remove for charring and the intermediate drying step can sometimes lead to deactivation of the plates, especially under humid atmospheric conditions, so that the second development does not produce the desired separations.

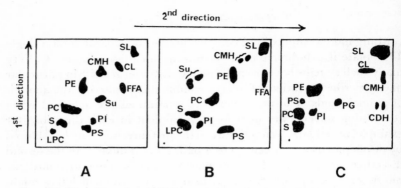

FIG. 7.3. Two-dimensional TLC separations of complex lipids from animal tissues on silica gel H containing 10 per cent (w/w) magnesium silicate (plates A and B) or silica gel H alone (plate C). Solvent systems: plate A,[446] 1st direction—chloroform–methanol–water (65:25:4 by vol.); 2nd direction—n-butanol–acetic acid–water (60:20:20 by vol.); plate B,[446] 1st direction—chloroform–methanol–28 per cent aqueous ammonia (65:35:5 by vol.); 2nd direction—chloroform–acetone–methanol–acetic acid–water (10:4:2:2:1 by vol.); plate C,[161] 1st direction—chloroform–methanol–water (65:25:4 by vol.); 2nd direction—tetrahydrofuran–methylal–methanol–2 M aqueous ammonia (10:5:5:1 by vol.). Abbreviations as in Fig. 7.1. (SL, simple lipids; FFA, free fatty acids).

Useful separations of plant lipids can also be achieved with two-dimensional TLC systems[374, 485] as shown in Fig. 7.4. Similar principles to those described above for animal lipids apply. Bacterial lipids are generally analysed in the same manner as plant lipids.

The amount of lipid that can be separated by two-dimensional TLC varies with the thickness of the layer of adsorbent. Up to 3 mg can be applied as a spot or a small streak (1 cm or so long) on layers 0·5 mm thick, but less material should be applied to layers 0·25 mm thick although better resolution is possible with the latter. The principle disadvantage of the technique lies in the fact that, although lipids migrate in a constant order in each direction, their precise orientation on the plate may vary considerably with small changes in the properties of the adsorbent. The double development is also time-consuming so that despite the lack of resolution, single-dimensional TLC systems are often preferred for routine analyses. For preliminary examination of lipid samples and for checks on the purity of lipid components isolated by other techniques, two-dimensional TLC offers considerable advantages.

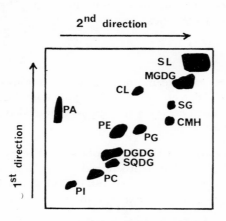

FIG. 7.4. Two-dimensional TLC separation of plant complex lipids[374] on layers of silica gel G. Solvent system, 1st direction—chloroform–methanol–7 N ammonium hydroxide (65:30:4 by vol.); 2nd direction—chloroform–methanol–acetic acid–water (170:25:25:6 by vol.). Abbreviations: SL, simple lipids; MGDG, monogalactosyl diglycerides; CL, cardiolipin; PA, phosphatidic acid; SG, sterol glycosides; CMH, ceramide monohexosides; PE, phosphatidyl ethanolamine; PG, phosphatidyl glycerol; DGDG, digalactosyl diglycerides; SQDG, sulphoquinovosyl diglyceride; PC, phosphatidyl choline; PI, phosphatidyl inositol.

3. *Location and identification of complex lipids on TLC plates*

Complex lipids can be located on TLC plates by most of the non-specific destructive or non-destructive reagents described in Chapter 3. If the latter are used, care must be taken to remove polar solvents such as water or glacial acetic acid in a vacuum oven at room temperature or by blowing nitrogen on to the surface of the plate otherwise lipid spots will be obscured. Acetic acid can also be removed by neutralisation with an ammonia spray (W. R. Morrison, personal communication). Such precautions are necessary to minimise autoxidation and patience and care at this stage are generally rewarded.

Iodine vapour can be used to distinguish between glycolipids and phospholipids as only the latter are stained significantly, but a number of more specific spray reagents are available that may be used to detect the presence of certain functional groups in complex lipids and, therefore, to aid in their identification.

(i) *Phosphorus*

The "Zinzadze" reagent in one or other of various modifications is commonly used to detect and identify lipids containing phosphorus. The following reagent can be especially recommended and has a shelf-life of several months.[552]

"Sodium molybdate (10 g) is dissolved in 4 N hydrochloric acid (100 ml) and hydrazine hydrochloride (1 g) is dissolved in water (100 ml). The solutions are mixed and heated on a boiling waterbath for 5 min. On cooling, the mixture is made up to 1 l when it is ready for use."

When TLC plates are sprayed with the reagent, phospholipids appear immediately as blue spots on a white background (this darkens after 1–2 hr). The better known but related spray of Dittmer and Lester[120] uses more corrosive reagents.

(ii) *Glycolipids*

As well as the rather negative use of iodine in conjunction with non-specific reagents to detect glycolipids (described above), a number of more positive highly-specific reagents are available for the detection of the carbohydrate moieties. That most widely used is an orcinol–sulphuric acid mixture.[526]

"The reagent is prepared by dissolving orcinol (200 mg) in 75 per cent sulphuric acid (100 ml). The whole surface of the plate is wetted by a fine spray of the solution and then is heated in an oven at 100°C for 10–15 min. Glycolipids appear as blue-violet spots on a white background."

The solution is stable for about one week if refrigerated and kept in the dark.

α-Naphthol[483] or diphenylamine[242] sprays also react specifically with glycolipids. The former reagent is prepared and used as follows.

"α-Naphthol (0·5 g) is dissolved in methanol–water (1:1, v/v) and sprayed on the plate until the surface is wet. After air drying, the plate is sprayed lightly with 95 per cent sulphuric acid then is heated at 120°C when glycolipids appear as purple-blue spots and other complex lipids as yellow spots."

(iii) *Compounds with vicinal diol groups*

All compounds of this type including phosphatidyl inositol, phosphatidyl glycerol, 1-monoglycerides and glycolipids can be detected with the periodate-Schiff's reagent.[479]

"Two reagents are necessary: (a) 0·2 per cent aqueous sodium periodate, (b) Schiff's reagent–pararosaniline (Fuchsin, 0·2 g) is dissolved in water (85 ml) and a 10 per cent solution of sodium bisulphite (5 ml) is added and the mixture allowed to stand overnight, when it is decolorised with charcoal and filtered. The TLC plate is sprayed with the periodate solution and left at room temperature for 15 min after which it is treated with sulphur dioxide to destroy excess reactant. The plate is then sprayed with the Schiff's reagent and treated once more with sulphur dioxide."

After a short time glycol-containing lipids appear as blue-purple spots, while most other complex lipids appear as yellow spots. Plasmalogens and free aldehydes are also detected by the procedure and the presence of such compounds should be separately determined by an independent analysis (see Chapter 6 and below). Unfortunately, autoxidised lipids also react so particular care must be taken to prevent the formation of lipid peroxides. When the glycolipids have been visualised, phospholipids can be detected on the same plate with the "Zinzadze" reagent.

(iv) *Free amino groups*

Phospholipids such as phosphatidyl ethanolamine, phosphatidyl serine and the related lyso compounds that have free amino groups can be detected with the aid of a ninhydrin spray.

"The plate is sprayed with a solution of 0·2 per cent ninhydrin in butanol saturated with water. Lipids having free amino groups appear as red-violet spots when the plate is heated in an oven at 100°C in a water-laden atmosphere."

(v) *Choline*

Phosphatidyl choline, lysophosphatidyl choline and sphingomyelin give a positive reaction with the "Dragendorff" reagent in the following modification.[566]

"Two solutions are necessary: (a) potassium iodide (40 g) in water (100 ml), (b) bismuth subnitrate (1·7 g) in 20 per cent acetic acid (100 ml). Immediately before use the first solution (5 ml) is mixed with the second solution (20 ml) and diluted with water (75 ml). When the plate is sprayed with the mixture, choline-containing lipids appear within a few minutes, especially if the plate is warmed gently, as orange-red spots."

The Levine–Chargaff test,[471] which is much more sensitive but less specific, is also used to detect the choline moiety on TLC plates. An alternative procedure, that is claimed to be more specific, has recently been described.[551]

(vi) *Sphingolipids*

The amide group of sphingolipids can be detected by means of a sodium hypochlorite ("Chlorox")–benzidine spray.[48, 492]

"Two reagents are necessary and both should be prepared immediately before use: (a) a saturated aqueous solution of sodium hypochlorite ("Chlorox" bleach, 5 ml) mixed with benzene (50 ml) and glacial acetic acid (5 ml); (b) benzidine dihydrochloride (200 mg; or benzidine itself) together with a small crystal of potassium iodide in 50 per cent ethanol (50 ml). The TLC plate is sprayed with the first solution and left at room temperature for 30 min. before it is dried at 80°C for 10 min. It is then sprayed with the second reagent— sphingolipids appear as blue spots."

(vii) *Gangliosides*

The sialic acid residues of gangliosides form a distinct and specific coloured complex with resorcinol.[528]

"A solution of resorcinol (2 g) in water (100 ml) is prepared and stored at 4°C. This stock solution (10 ml) is added to concentrated hydrochloric acid (80 ml) and 0·1 M copper sulphate solution (0·25 ml) and made up to 100 ml immediately before use. The plate is sprayed with this reagent and heated at 110°C for a few min. A violet-blue colour is obtained with gangliosides while other glyco-lipids appear as yellow spots."

(viii) *Identification of phospholipids by IR spectroscopy*

The nitrogenous bases of phospholipids can be identified by their IR absorption spectra in the regions 9–11 μm.[370] Free amine groups, such as are found in phosphatidyl serine or phosphatidyl ethanolamine, exhibit a single sharp band with a maximum at 9·3 μm while mono-methylamines (e.g. in phosphatidyl N-methylethanolamine) display a strong maximum at 9·5 μm with a weaker band at 9·2 μm. The spectra of dimethylamines exhibit a pronounced doublet with maxima at 9·2 and 9·5 μm while quaternary amines, such as choline in phos-phatidyl choline or sphingomyelin, have in addition to this doublet a sharp band at 10·4 μm. In addition, the spectra of these compounds in the IR region 5·5 to 7·0 μm also exhibit features useful as diagnostic aids.[36] Spectra are normally obtained with the phospholipids in chloroform solution.

D. Determination of Complex Lipids Separated by
Chromatographic Procedures

No single method has become established as the standard for estimation of complex lipids separated by chromatographic procedures, but a variety of methods are available for use in differing circumstances. Universal procedures such as charring techniques[408] or fluorometry[201] are little used for complex lipids as standards are not always available for preparing calibration curves, but they may be suitable for screening large numbers of similar samples. Moving wire flame ionisation detection systems have yet to be widely applied in column chromatography of complex lipids.

Phospholipids separated chromatographically can be estimated by determining the amount of phosphorus in each compound after digestion to inorganic phosphate and a detailed procedure suitable for the purpose is described below. Components should preferably be eluted from TLC adsorbents with chloroform–methanol (1:1, v/v) containing 10 per cent water after detection with a 2′,7′-dichlorofluorescein spray or with iodine vapour although the estimation can be carried out in the presence of the adsorbent if necessary. If this is not washed thoroughly with solvent before the plates are prepared (see Chapter 3), however, the blank values may be high and the accuracy of the procedure diminished. In calculating molar amounts of phospholipids from phosphorus analyses, it should be remembered that certain of these, e.g. cardiolipin, contain 2 moles of phosphorus per mole of total lipid. One disadvantage of the technique is that it is destructive and when small amounts only of material are available for analysis, it may not be possible to obtain any more information from the sample.

Glycerophosphatides and glycosyl diglycerides can also be determined by estimation of the glycerol moiety after acetylation,[408] but full details of the procedure have yet to be published.

Glycosphingolipids separated by chromatographic means are generally estimated by disrupting the glycosidic and amide bonds with aqueous acid and isolating the sphingosine bases, the amounts of which are then determined from an analysis of the nitrogen present in the compounds.[146]

"The lipids (up to 4 mg) are heated under reflux for 3 hr in 2 N methanolic hydrogen chloride when the mixture is taken to dryness. 3 N Aqueous hydrochloric acid (1 ml) is then added and the products are heated in a stoppered tube at 100°C for 1 hr. On cooling, chloroform–methanol (2:1, v/v; 1 ml) is added, the aqueous layer, which contains the liberated carbohydrates, is removed carefully by means of a Pasteur pipette and the chloroform layer is washed twice with fresh upper phase (prepared by partitioning chloroform–methanol (2:1, v/v) with an equal volume of water). The chloroform layer contains only long-chain bases and free fatty acids and the amount of the former can be determined by nitrogen analysis by standard micro-Kjeldahl procedures."

Alternatively the bases can be determined spectrophotometrically and with greater sensitivity as the trinitrobenzenesulphonic acid derivatives.[606]

"The chloroform extract obtained as above is taken to dryness and 4 per cent aqueous sodium bicarbonate solution (1 ml) and 1 per cent aqueous trinitrobenzenesulphonic acid solution (1 ml) are added and the mixture is left in the dark for 1 hr at 40°C. Methanolic hydrogen chloride (1 N, 1 ml) is added and the solution extracted three times with hexane (2 ml portions). The hexane is removed and the sample dissolved in 95 per cent ethanol (4 ml) when the absorbance of the solution at 340 nm is obtained relative to an appropriate blank. The amount of glycolipid is read from a calibration curve prepared with pure standards."

Lipids containing free amine groups, such as phosphatidyl ethanolamine or phosphatidyl serine, can be determined by a related procedure.[482] Glycolipids can also be determined by analysis of the carbohydrate residues (after reaction with anthrone),[606] but as the molar proportions of these in a given lipid may not be known, the procedure is not definitive (see p. 226).

It is also possible to determine the amounts of complex lipids, particularly those separated by TLC, by gas chromatography of the methyl ester derivatives of the fatty acid constituents with an added internal standard.[90, 99] The amounts and fatty acid compositions of all lipid classes are thereby obtained simultaneously in a single analysis. As fatty acids are the only hydrolysis product common to all glycero-phosphatides and glycolipids, the method is particularly apposite. The principle of the procedure has been described in Chapter 3 and the application of the method to the analysis of simple lipids in Chapter 6. Glycerophosphatides can be transesterified by alkali-catalysed methanolysis in the presence of the TLC adsorbent (see Chapter 4), but sphingolipids must be transesterified by methanol containing hydrochloric acid (see Chapter 4 and below). If the internal standard (15:0 or 17:0) in methanol solution (about 0·2 mg/ml) is added to the lipid and the adsorbent along with the transesterifying reagent and the reaction is taken to completion, there will be virtually no loss of lipid relative to the internal standard. The procedure has been tested with a wide range of phospholipids, including sphingomyelin, but not with glycolipids although it is potentially of use here also. Again, in calculating the molar proportion of each lipid class, it must be remembered that some contain only one fatty acid per molecule and others two or more.

The determination of individual hydrolysis products from complex lipids is discussed in greater detail below.

E. Analysis of Complex Lipids as Their Partial Hydrolysis Products

Methods of estimating glycerophosphatides after deacylation have been developed in a number of laboratories and are especially advocated by Dawson,[111] who was the principal originator of the procedure. Lipids are deacylated by treatment with mild alkali and the water-soluble phosphorus-containing compounds are isolated and separated by paper chromatography. These are more easily separated by this technique than are the parent phosphatides by adsorption TLC and, in particular, the acidic phospholipids are especially well-resolved. In addition, ether and vinyl ether bonds are not disrupted by the procedure so that these forms of each lipid class can also be separated and estimated. A great deal of information is therefore obtained that is otherwise not readily available. The procedure has been recommended for lipids labelled isotopically in the glycerol or the phosphorus moiety.

The principal disadvantages of the technique are that no information can be obtained about the fatty acids attached to each phosphatide and that lysophosphatides (although they are usually comparatively minor components of tissues) cannot be distinguished from the diacyl compounds. It is also a somewhat lengthy and tedious procedure which must be performed under rigorously controlled conditions and it is necessary to make certain corrections for side reactions. Somewhat unfairly, it is shunned by lipid chemists who tend to eschew the handling of organic compounds in aqueous media if at all possible. This hydrophobia would no doubt be overcome if a procedure were developed for complementary analysis of the intact lipids as well as the deacylated products because of the extra information that might be obtainable.

F. Sphingolipids

As sphingolipids, with the exception of sphingomyelin, are only trace components of most tissues, it is usual to separate glycosphingolipids from phospholipids and to analyse the former separately. In addition, as the hydrolysis products tend to be very different from those obtained with phospholipids, the chemical problems involved in their analysis are not the same. Methods of analysis of glycosphingolipids are especially important because of the association of these compounds with certain physiological and pathological disease states.

1. *Cerebrosides and ceramide polyhexosides*

Cerebrosides and ceramide polyhexosides as a class are obtained as described above by elution with acetone or tetrahydrofuran–methylal–methanol–water (10:6:4:1 by vol.) from silicic acid columns or with superdry solvents from Florisil. In addition, they are obtained in specific fractions from DEAE and TEAE cellulose columns.

Chemical procedures can also be used. The complex lipids are subjected to mild alkaline transesterification by means of which those lipids containing O-acyl bound fatty acids are converted to methyl esters and water-soluble products while sphingomyelin and other glycosphingolipids are unaffected. The following method is suitable:

"The complex lipids (100 mg) are heated at 50°C for 10 min in 0·5 M sodium methoxide in methanol (3 ml). Acetic acid (0·15 ml) is added followed by chloroform (6 ml) and water (2 ml). After thorough shaking, the upper layer is removed by Pasteur pipette and the lower layer is washed twice more with methanol–water (1:1, v/v; 2 ml portions). After removal of the solvent, the crude products must be purified by chromatography."

The hydrolysis products of the glycerophosphatides, including any deacylated ether derivatives, are easily removed by column chromatography and if need be the glycosphingolipids can be separated from sphingomyelin. Chloroform elutes the methyl esters then chloroform–methanol (1:1 by vol.) the pure glycosphingolipids from silicic acid columns.[154] With animal lipids, the acetone fraction from silicic acid columns should also be treated in this way to eliminate traces of the acidic phospholipids, although traces of glycosyl diglycerides that may be present in some tissues will also be removed by such a step.

More thorough separations of glycolipids from phospholipids can be obtained if the carbohydrate moieties of the former are first acetylated with pyridine–acetic anhydride (see Chapter 4 for practical details).[455] Acetylated total lipid extracts are chromatographed on Florisil; 1,2-dichloroethane elutes neutral lipids while 1,2-dichloroethane–acetone (1:1, v/v) elutes all the acetylated glycolipids completely free of phospholipids. The latter remain on the column from which they can be recovered (although yields may be poor) if necessary. The glycolipids are deacetylated with sodium methoxide in methanol as above before they are analysed further. Gangliosides are also obtained in the glycolipid fraction if crude unwashed lipid extracts are taken to dryness, acetylated and fractionated in this way, although the deacetylated product must later be purified by dialysis against ice-water to remove non-lipid contaminants. No detectable structural changes occur in the process.

Two-dimensional TLC systems that separate individual glycolipids along with phospholipids have been described above. When phospholipids are absent, glycolipids are easily resolved by TLC in a single dimension. For example, Svennerholm and Svennerholm[525] separated four main classes of glycolipids, i.e. cerebrosides, ceramide di- and trihexosides and ceramide trihexoside–N-acetyl-galactosamine, on silica

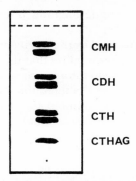

Fig. 7.5. TLC separation of glycosphingolipids on layers of silica gel G.[525] Solvent system, chloroform–methanol–water (65:25:4 by vol.). Abbreviations: CMH, CDH and CTH, ceramide mono-, di- and trihexosides respectively; CTHAG, ceramide trihexoside–N–acetylgalactosamine.

gel G layers with solvent system chloroform–methanol–water (65:25:4 by vol.) as illustrated in Fig. 7.5. In addition, the first three of these components appear as double bands as those molecules containing normal fatty acids are separated to some extent from those containing hydroxy fatty acids.[158, 525] In fact, the separation is not as complete as it is sometimes claimed to be as under optimum conditions, three bands should be separable, i.e. fractions containing normal fatty acids with dihydroxy bases, hydroxy fatty acids with dihydroxy bases together with normal fatty acids with trihydroxy bases, and finally hydroxy fatty acids with trihydroxy bases. The pattern can occasionally be more complicated as compounds containing very long-chain fatty acids are often partially separated from those having a more normal range of chain lengths;[221] sphingomyelin frequently chromatographs as a double spot for the same reason.[586] Cerebroside sulphate migrates just ahead of ceramide dihexosides with the above TLC system. Glycolipids containing four to seven hexose units are not easily separated unless they are first acetylated.[179, 607]

The nature of the hexose unit affects the chromatographic behaviour of glycolipids on silica gel only slightly, but ceramide galactoside is clearly separated from ceramide glucoside on silica gel impregnated with sodium tetraborate (approx. 5 per cent w/w). With the solvent system used above, the glucoside migrates just ahead of the galactoside.[608]

Glycolipids may be recognised by the specific spray reagents described earlier, but a complete analysis requires that the molar amounts of each hexose be determined relative to the amount of ceramide and that the fatty acid and long-chain base compositions be ascertained by the methods discussed below. In addition, the sequence of hexose units should also be determined, but this cannot be discussed here.

2. *Gangliosides*

Gangliosides, although they are only minor components of most tissues, are becoming increasingly the subject of study because of their role in various lipidoses. They are found in the aqueous phase along with non-lipid contaminants after a "Folch" wash of lipid extracts and they can be recovered from this, after dialysis, by lyophilisation (discussed earlier in Chapter 2).[248, 323] Alternatively, a pure ganglioside fraction can be obtained by Sephadex chromatography[483] or they can be obtained along with other glycosphingolipids by Florisil chromatography of acetylated lipid extracts (as discussed above).[455]

Individual gangliosides can then be isolated by preparative TLC in a single dimension, but very long development times may be necessary because of the highly polar nature of the compounds. Two groups of developing solvents have proved useful—basic, e.g. chloroform–methanol–2.5 N aqueous ammonia (60:40:9 by vol.),[301] or neutral, e.g. propanol–water (70:30, v/v).[282] Acidic solvent systems must be avoided as they may cause degradation of gangliosides. The purity of fractions isolated with one type of solvent system should be checked by the alternative type as certain components which co-chromatograph with one solvent system are separable with another of a different type. Figure 7.6 illustrates the kind of separations that are attainable. Penick *et al.*,[396] in particular, have studied TLC systems for the separation of gangliosides in some detail. The resorcinol spray described above can be used to identify them on TLC plates.

Individual gangliosides can be recognised partly by their *Rf* values in various solvent systems relative to compounds isolated from tissues of

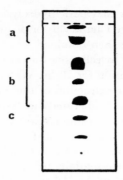

FIG. 7.6. TLC separation of gangliosides on layers of silica gel G. Solvent system, chloroform–methanol–2·5 N ammonia (60:40:9 by vol.). a, mono-, b, di-, and c, trisialogangliosides.[299]

known ganglioside composition, e.g. bovine brain.[300] More positive identification requires that the hexose and sialic acid residues be identified and the molar amounts of each determined as well as the fatty acid and long-chain base compositions. Methods of analysing for all these components are discussed below. Complete identification requires that the mode of linkage of each of the hexose and sialic acid residues be ascertained, but structural studies of this kind are out with the scope of this text. The chemistry of gangliosides has been reviewed.[300, 323]

3. *Long-chain bases*

(i) *Isolation of bases*

In order to obtain the long-chain bases from sphingolipids, it is first necessary to hydrolyse any glycosidic linkages in glycolipids or the phosphate bond in sphingomyelin as well as the amide bond before the various products are isolated and purified. The chemistry and methods of analysis of long-chain bases have been reviewed by Karlsson.[254]

Although the amide and phosphate bonds in sphingomyelin and ceramide aminoethylphosphonate can be cleaved simultaneously by acidic hydrolysis, better results are achieved if the compounds are first converted to ceramide by enzymatic hydrolysis with the phospholipase C of *Clostridium welchii* (see Chapter 9) prior to hydrolysis of the amide bond. The free bases are then obtained by alkaline hydrolysis with a minimum of degradation by refluxing the ceramides in 1 M methanolic potassium hydroxide solution for 18 hr. Water is added and the bases obtained by diethyl ether extraction of the alkaline solution (see also Chapter 4).[364] The fatty acids liberated can also be recovered if necessary by acidifying the aqueous layer and re-extracting. A related procedure in which the hydrolysis products are purified by a rapid column method has been described by Karlsson.[253] Long-chain bases prepared in this way via the ceramides are essentially free of degradation products.

Ceramides are less easily prepared from glycosphingolipids unfortunately. Compounds containing dihydroxy bases can, however, be converted to ceramides by an elegant procedure devised by Carter *et al.*[76] in which the glycosidic ring is opened with periodate and the resulting product reduced with sodium borohydride before it is hydrolysed under mild acidic conditions. Derivatives of trihydroxy bases cannot be converted to ceramides by this procedure as the bases would be cleaved across the vicinal diol group by the periodate oxidation step. Indeed, the preparation of ceramides or long-chain bases in the pure state from glycolipids is one of the more important unsolved technical problems in lipid methodology. Although a number of enzymes have been described that will hydrolyse carbohydrate–ceramide bonds,[354] they do not appear to have been used in structural studies possibly because the reactions do not go to completion. In particular, preparations of β-glucosidase (EC. 3.2.1.21) that release ceramide from glucosyl ceramide are easily obtained.[148] Instead it is more usual for the sugar and fatty acid moieties to be removed from the long-chain bases simultaneously by acidic hydrolysis procedures although during this process considerable degradation and modification of the bases, particularly those which contain allylic double bonds, occurs. For example, 3- or 5-methoxy and 5-hydroxy-compounds are formed, some dehydration may occur and there is liable to be some inversion of configuration of the hydroxyl and amine groups. The most widely used acidic hydrolysis procedure is that of Gaver and Sweeley[150] in which by-product formation is less than in other methods, although the trihydroxy-bases are considerably altered.

"The acidic reagent is prepared by mixing concentrated hydrochloric acid (8·6 ml) and water (9·4 ml) and making up to 100 ml with methanol. The glycolipid (up to 4 mg) is heated with the reagent (2 ml) in a sealed tube at 70°C for 18 hr. The mixture is then taken to dryness in a stream of nitrogen and the bases are isolated by ion-exchange chromatography or, after neutral and acidic materials have been removed by diethyl ether extraction of an aqueous acid solution, by extracting with diethyl ether from aqueous alkali."

Most of the amide-linked fatty acids are converted to methyl esters during the reaction and after purification by TLC, they can be analysed by gas chromatography (see below). The main by-products formed from the bases are 5-methoxy compounds. Karlsson[254] does not favour this method when analysing for trace components as methoxy artefacts are not easily separated from the native bases. Rather, he recommends the following procedure.

"The glycolipids (4 ml) are refluxed for 6 hr with 2 M aqueous hydrochloric acid (3 ml). On cooling, chloroform (8 ml) and methanol (4 ml) are added and, after thorough mixing, the lower phase is evaporated to dryness and the products are added in chloroform to a silicic acid column (1 g). Chloroform–methanol (98:2, v/v; 10 ml) elutes the free fatty acids and chloroform–methanol (1:3, v/v; 10 ml) elutes the bases."

By-product formation is greater in this instance than with the Gaver and Sweeley[530] procedure, but 5-hydroxy compounds are formed rather than methoxy compounds and the former are more easily separated from the natural bases, particularly when the amino-group is stabilised by conversion to the dinitrophenyl (DNP) derivative as described below.[254]

"The bases (5 mg) are dissolved in methanol (1 ml) containing 1-fluoro-2, 4-dinitrobenzene (5 μl). 2 M Potassium borate buffer (pH 10·5, 4 ml) is added dropwise and the solution is heated at 60°C for 30 min. On cooling, the mixture is partitioned between chloroform, methanol and water in the ratio 8:4:3 by vol. and the lower phase is evaporated and the material chromatographed on a silicic acid column (1 g). Hexane–diethyl ether (7:3, v/v; 20 ml) elutes the less polar impurities, hexane–ether (1:1, v/v; 20 ml) elutes the DNP

H

derivatives of the natural bases and hexane–ether (1:3, v/v) elutes more polar artefacts."

The bases can be further analysed in the form of the non-polar DNP derivatives which are yellow in colour and are easily seen on a TLC plate.

(ii) *Separation of individual long-chain bases and their derivatives*

DNP derivatives of long-chain bases are readily separated by TLC on layers of silica gel G impregnated with 2 per cent (w/w) boric acid into three groups. i.e. saturated dihydroxy-, unsaturated dihydroxy- (with a *trans*-double bond in position 4) and trihydroxy- components, with chloroform–hexane–methanol (50:50:14 by vol.) as developing solvent as illustrated in Fig. 7.7.[254, 256] When the bases are prepared by acidic hydrolysis methods, unnatural *threo*-isomers of the unsaturated dihydroxy- bases are found just below the natural *erythro*- compounds on the TLC plate, but they need not interfere with subsequent GLC analyses. Components are recovered from the adsorbent by elution with chloroform–methanol (2:1, v/v), but the eluate should be washed with ⅕ volume of water to remove any boric acid which may also be eluted.

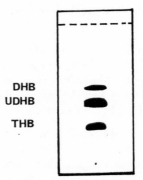

DHB
UDHB
THB

Fig. 7.7. TLC separation of dinitrophenyl-derivatives of long-chain bases on layers of silica gel G containing 2 per cent (w/w) boric acid.[254, 256] Solvent system, chloroform–hexane–methanol (50:50:14 by vol.). Abbreviations: DHB, saturated dihydroxy bases; UDHB, $\triangle^4$-*trans*-unsaturated dihydroxy-bases; THB, trihydroxy bases.

DNP derivatives of bases separated in this way can be further resolved according to degree of unsaturation by silver nitrate chromatography[252] or into critical pairs by reverse phase TLC.[250] By using these techniques in sequence, individual components of natural mixtures can be isolated in sufficient quantity for structural studies.

The free bases or other derivatives of the bases can also be separated by TLC although the resolutions attainable are not quite as good as those with the more stable DNP derivatives,[254, 572] but as steps involving preparation and hydrolysis of derivatives are avoided, many workers prefer to work with the free bases.

Isomers of the various classes of bases can also be separated by gas chromatography if they are first converted to non-polar derivatives, in particular to the trimethylsilyl ethers. Components differing in chain length or in the number of hydroxyl groups in the molecule can then be separated on silicone liquid phases such as SE-30.[150] The N-acetyl, O-TMS ether derivatives of long-chain bases are also useful for gas chromatography. The N-acetyl derivative is prepared by reacting the free base with acetic anhydride in methanol (1:4, v/v) at room temperature overnight (under these conditions no O-acetylation occurs).[151] The TMS ether derivatives are prepared from the N-acetyl compound as described earlier (Chapter 4) and may be subjected to GLC on silicone liquid phases such as OV-1 or OV-17 (3 per cent stationary phase at column temperatures of 220 to 240°C). Although isomers differing in degree of unsaturation cannot be separated on such columns, this can be achieved if the double bonds are first hydroxylated and then converted to the TMS ether derivatives (see Chapter 4 for practical details).

The facility to chromatograph intact long-chain base derivatives is especially important when they are isotopically labelled, but much better resolutions are obtained if the bases or their DNP derivatives are oxidised by periodate to form long-chain aldehydes. Such compounds are readily separated both by chain-length and degree of unsaturation by GLC on conventional polyester liquid phases as discussed in Chapter 5. Di- and trihydroxy bases, whether they are of the *threo-* or *erythro-* configuration, give similar aldehydes (although aldehydes of trihydroxy bases will be one carbon atom shorter than those of dihydroxy bases) so it is essential to isolate the three major classes of bases by preparative TLC as described above prior to GLC analysis. Various procedures utilising sodium metaperiodate have been described for cleaving the bases, but the mildest reagent is probably sodium metaperiodate in methanol[530] used as follows.

"0·2 M aqueous sodium metaperiodate (1 ml) is added to the long-chain bases (5 mg) in methanol (5 ml). The mixture is stirred in the dark at room temperature for 1 hour when dichloromethane (10 ml) and water (5 ml) are added and the whole is shaken and centrifuged. The aqueous layer is extracted twice more with dichloromethane and the combined extracts are dried over anhydrous sodium sulphate before the solvent is removed."

The aldehydes derived from bases are more easily identified than the parent compounds as a wide range of the former are available for use as standards. Aldehydes can also be reduced to long-chain alcohols with sodium borohydride and converted to the TMS ether or acetate derivatives for GLC analysis[364] and this is preferred by some workers.

When bases are prepared by acidic hydrolysis, it should be remembered that, although the relative amounts of the various isomers from each class of base determined by GLC analysis reflect reasonably accurately the amounts present in each class in the native state, the amounts of the three groups of bases relative to each other do not. Saturated dihydroxy-bases are not affected by the hydrolysis reaction, but as little as one-half the original amount of the allylic bases may be recovered and there will also be losses of trihydroxy bases. As stated earlier, no method of circumventing this problem is yet available.

(iii) *Identification of long-chain bases*

Karlsson[255] has reviewed procedures for determining the structures of long-chain bases isolated by the above procedures. The most powerful tool for identifying them is mass spectrometry. The natural bases are generally converted to less polar derivatives for the purpose; for example, Karlsson[254] converts the bases to the DNP derivatives before methylating the free hydroxyl groups by refluxing them overnight with methyl iodide and silver oxide. These derivatives give characteristic fragmentation patterns in a mass spectrometer so that the molecular weight and primary structure of the base can be ascertained. The positions of double bonds in the chain can also be established if they are first hydroxylated and then converted to the TMS ethers in a manner similar to that discussed earlier for unsaturated fatty acids (see Chapter 5).

Definitive mass spectra are also obtained from N-acetyl, O-TMS ether derivatives of long-chain bases (prepared as described above).[400, 401] In addition, these compounds are sufficiently volatile for GLC-mass spectrometer combinations to be used for the identification of components of natural mixtures.

The long-chain aldehydes produced by periodate oxidation of the bases or alcohols prepared from the aldehydes can also be identified by mass spectrometry as well as by their GLC retention characteristics and the positions of any double bonds can be established by permanganate–periodate oxidation (see Chapter 5 for a description of the method) with GLC identification of the fatty acid fragments.[364] The number of hydroxyl groups in the intact molecule cannot of course be obtained if the aldehydes alone are examined by this technique, but this feature can often be established from the chromatographic behaviour of the intact bases or their derivatives on thin-layer adsorbents or by gas chromatography if authentic standards are available.

Infrared absorption spectroscopy can be used to advantage to detect the presence of *trans* double bonds in the aliphatic chain of a long-chain base or its derivative by means of the characteristic absorption band at $10\cdot3\ \mu\text{m}$.

(iv) *Fatty acids associated with long-chain bases*

Although the problems of analysis of fatty acids have been discussed at length in Chapter 5, a reminder may be appropriate that sphingolipids contain two unique groups of fatty acid constituents—very long-chain odd and even numbered saturated and monoenoic fatty acids (up to C_{28}) and a group of related compounds with hydroxyl groups in position 2, both of which are linked by amide bonds to the long-chain bases. In addition, there may occasionally be fatty acids linked to hydroxyl groups also. The problems involved in the preparation of methyl esters from the above have been discussed in Chapter 4; O-acyl esters are transesterified with sodium methoxide and N-acyl with methanol containing hydrochloric acid.

If the original sphingolipids have not already been separated by TLC according to which group of fatty acids they contain, the methyl esters of the two types of fatty acid are readily separated by TLC (see Chapter 5). Alternatively, as 2-hydroxy fatty acids form copper chelates, they can be isolated as such.[274] Normal fatty acids are identified by the standard procedures described earlier (Chapter 5). The hydroxy esters can be analysed by GLC after conversion to the acetate or TMS ether derivatives and it should also be possible to analyse the aldehydes formed by periodate cleavage between the hydroxyl and carboxyl groups by GLC.

G. Alkyl- and Alkenyl-ether Derivatives of Complex Lipids

Although simple alkyl ethers and plasmalogens can be separated with care from the analogous acyl derivatives (see Chapter 6), related phosphoglycerides of the three types cannot be separated in the native form by any procedure yet devised. Phospholipids isolated by all the procedures described above may, therefore, include alkyl–acyl- and alkenyl–acyl- forms of the glycerophosphatide in addition to the more common diacyl derivatives. While such ether-containing lipids are not often major components of tissues (other than heart muscle or nervous tissue) they are almost always present in trace amounts.

The analysis of the various hydrolysis products of simple plasmalogens and alkyl ethers, principally aldehydes and 1-alkoxy-glycerols respectively, has been discussed in Chapter 6 and similar procedures are used to detect these moieties in complex lipids. For example, plasmalogens can be detected by spraying TLC plates with the 2,4-dinitrophenyl hydrazine reagent described earlier; aldehydes are released by the action of aqueous acid or of mercuric chloride and are analysed as described in Chapter 6. 1-Alkoxy-glycerols and 1-alkenyloxy-glycerols are liberated from phosphoglycerides by hydrogenolysis with lithium aluminium hydride[596] and are also analysed as described earlier.

Plasmalogens can in addition be recognised and determined by reaction TLC.[391] The complex lipids (about 1 mg) are applied to the bottom left-hand corner of a TLC plate that is developed in chloroform-methanol–water–acetic acid (65:43:3:1 by vol.). Volatile solvents are removed from the plate in a stream of nitrogen, residual acetic acid is neutralised by exposing the plate to ammonia vapour for some time and excess ammonia is removed in a desiccator or vacuum oven at room temperature before the lipid lane is sprayed with 5 mM aqueous mercuric chloride solution. After 1 min, the plate is turned 90° anticlockwise and developed in chloroform–methanol–water (60:35:8 by vol.). The spots are visualised by standard procedures and results similar to that illustrated in Fig. 7.8 are obtained. The aldehydes liberated along with the lysophosphatides formed simultaneously are clearly separated from the unchanged diacyl- and alkyl–acyl-phosphatides so that, if non-destructive spray detection reagents are used, the aldehyde and fatty acid components of the plasmalogens can be obtained and analysed while the relative amounts of related plasmalogens and non-plasmalogenic components are determined by phosphorus analyses of the appropriate spots.

Unfortunately none of these procedures permit the identification of the fatty acids associated with either the alkyl–acyl- or diacyl-phospholipids when both forms of the lipid are present in the sample and in

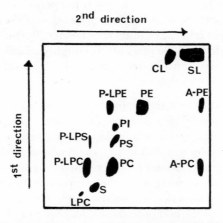

FIG. 7.8. Two-dimensional reaction TLC of phospholipids[391] on silica gel H layers. Solvent system, 1st direction—chloroform–methanol–water–acetic acid (65:43:3:1, by vol.). Lipid lane is then sprayed with 5 mM mercuric chloride. Solvent system, 2nd direction—chloroform–methanol–water (60:35:8, by vol.). Abbreviations: SL, simple lipids; CL, cardiolipin; PE, phosphatidyl ethanolamine; PI, phosphatidyl inositol; PS, phosphatidyl serine; PC, phosphatidyl choline; S, sphingomyelin; LPC, lysophosphatidyl choline; prefix P-L and A-, lysophosphatide and aldehydes respectively derived from the appropriate plasmalogen.

addition, the specific activity of ^{32}P in the various structurally related compounds, cannot be determined in isotope-turnover experiments. These problems can sometimes be solved by modifying lipids chemically or enzymatically.

Concentrates of plasmalogens and alkyl-phosphatides can be prepared by treating the mixed phospholipids briefly with mild alkali; the diacyl forms of the phospholipids are transesterified much more rapidly than are the ether-containing analogues and concentrates of the latter are obtained. Pure alkyl-acyl-phosphatides can then be prepared by selectively hydrolysing the plasmalogens with acid.[423, 435] The procedure has been used successfully with phosphatidyl cholines,[423, 435] phosphatidyl ethanolamines [424] and phosphatidyl N-methyl ethanolamines.[102] The following method is recommended by Renkonen.[430]

"The lipid (3 mg) is dissolved in chloroform–methanol (1:1, v/v; 1 ml) and 0·35 N sodium hydroxide in 96 per cent methanol (0·16 ml) is added. The mixture is left at room temperature for up to 25 min (determined experimentally and depends on how much ether-containing lipid is in the sample) when water (0·5 ml) and chloroform (1 ml) are added and the reagents thoroughly mixed. The lower layer is washed once with water–methanol (1:1, v/v; 1 ml) and dried over anhydrous sodium sulphate."

The required ether-containing lipids are separated from the partial hydrolysis products by chromatographic means. The procedure suffers from the disadvantage that a considerable proportion of the alkyl- and alkenyl-phosphatides are also lost in the reaction so that the amounts of these in the sample cannot be determined. It appears, however, that there is no fatty acid selectivity in the components lost so that the alkyl-phosphatides or plasmalogens obtained are representative in nature if not in amount of the compounds originally present in the sample and they can be used for determinations of molecular species.[430]

Concentrates of ether-containing lipids can also be obtained by enzymatic hydrolysis of phosphatides. Phospholipase A,[157] phospholipase C[568] and phospholipase D[296] (see Chapter 9 for an extensive discussion of these enzymes) all may react preferentially, but not exclusively, with diacyl phosphatides under certain conditions and concentrates of ether-containing compounds are obtained if the reaction is not permitted to proceed too far. A more promising approach is to hydrolyse the phosphatides with highly purified pig pancreatic lipase (see also Chapter 9) which removes only those fatty acids in position 1 of diacyl phosphatides while those compounds that contain fatty acids in position 2 only are not affected.[433] Unfortunately, pure enzyme preparations are not readily available and crude commercial enzyme extracts may not be suitable as they usually contain phospholipases which react, albeit slowly, with ether-containing lipids so that recovery of these is not complete.

None of the above procedures permit the quantitative separation of intact diacyl-, alkyl–acyl-, and alkenyl–acyl- forms of phosphatides from each other. This may be possible, however, if the phospholipids are first converted to non-polar derivatives. For example, phosphatidyl choline can be reacted with phospholipase D with the formation of phosphatidic acid which is in turn converted to a non-polar dimethylphosphatide by reaction with diazomethane (see Chapter 9 for further practical details). By means of TLC on silica gel layers with diethyl ether or hexane–chloroform (1:1, v/v) as developing solvent and repeatedly running the plate in the same direction, this compound can be separated into alkenyl-, alkyl- and diacyl-forms.[432, 433] The alkenyl groups may form a by-product with diazomethane unfortunately, which limits the potential of the method. Phosphatidyl ethanolamines have been converted to the N-dinitrophenyl, O-methyl derivative which is sufficiently non-polar to be separable into the three groups of components by TLC.[432] The derivatives are prepared as follows:

"The phosphatidyl ethanolamine (1 mg) is dissolved in benzene (2 ml) and triethylamine (40 μl) and fluoro-2,4-dinitrobenzene (4μl) are added. After 2 hr the reagents are removed *in vacuo* (< 1 mm at 90°C). The product in chloroform (1 ml) is methylated with excess diazomethane in diethyl ether at 20°C for 30 min."

Repeated TLC on silica gel G layers in a single dimension (up to 8 times) is necessary to separate effectively the diacyl and ether-containing forms of the compound with hexane–chloroform (4:6, v/v) and/or toluene–chloroform (4:6, v/v) as solvent systems. The principal advantage of a procedure of this type is that the phosphorus atom is retained and metabolic studies of the various subgroups of phospholipids with ^{32}P as marker are possible.

The three subgroups of each phospholipid class can also be separated if the polar head group is removed by reaction with phospholipase C (see Chapter 9 for practical details) and the resulting partial glycerides rendered less polar by acetylation. Alkenyl–acyl-, alkyl-acyl-, and diacyl–glycerol acetates prepared in this way have similar mobilities on thin layers of silica gel G to the related simple lipids described earlier (Chapter 6). Excellent separations are possible if the plate is developed first half-way with hexane–diethyl ether (1:1, v/v) as solvent and then fully in the same direction with toluene.[436, 437]

To summarise, for the analysis of the three subgroups of natural phosphatides, it is first necessary to determine the precise amount of each by a hydrolytic technique of some kind, for example by the reaction with lithium aluminium hydride in which 1-alkyl- and 1-alkenyl-glycerols are obtained for estimation (see Chapter 6). Although it has not been discussed in detail here, the hydrolysis procedure of Dawson[111] is also excellent for the purpose. Individual components can then be isolated for more detailed analyses by one of the procedures above which guarantee recovery of representative samples of each lipid type, although the overall yields may not be good. Useful reviews on the subject have appeared elsewhere.[430, 559]

H. Phosphonolipids

Phosphonolipids are not easily separated from the related phospholipids as they have very similar polarities and they are generally isolated together, the presence of the former being detected mainly by the recovery of hydrolysis products containing a carbon–phosphorus bond that is stable to prolonged acidic hydrolysis. Ceramide aminoethylphosphonate, for example, is eluted with phosphatidyl ethanolamine from DEAE cellulose columns and most other phosphonate lipids migrate with their phosphoryl analogues with the more usual TLC systems. Kapoulas,[249] however, has described chloroform–acetic acid–water solvent systems which will separate phosphatidyl choline and phosphatidyl ethanolamine and their phosphonate analogues. For example, with the above solvents in the ratio 60:34:6 (by vol.), phosphonate components migrate ahead of the more common phosphate compounds as illustrated in Fig. 7.9. Berger and Hanahan[45] have also used chloroform–acetic acid elution mixtures to separate the natural phosphonate derivative of phosphatidyl ethanolamine from its phosphoryl analogue on silicic acid columns. Before such separations are attempted, it is probably advisable to isolate groups of related components from natural mixture by conventional procedures.

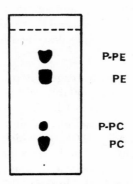

FIG. 7.9. TLC separation of phospho- and phosphono- analogues of phospholipids on silica gel G;[249] solvent system, chloroform–acetic acid–water (60:34:6 by vol.). Abbreviations: PC, phosphatidyl choline; PE, phosphatidyl ethanolamine; prefix "P", phosphono- analogue of each phosphatide.

Phosphonate bases are liberated from phosphonolipids by acidic hydrolysis and a variety of chromatographic methods for the isolation and identification of these have been described of which the most useful are probably paper chromatography[113] and TLC.[510] Snyder and Law[510] have described a procedure for the determination of phosphonate phosphorus which offers a number of advantages over others available. Lipids are digested first with acid and then by reaction with the enzyme alkaline phosphatase (from *E. coli*) under conditions such that all the phosphorus in phosphate form is degraded to inorganic phosphate although the phosphorus–carbon bond is not disrupted. The inorganic phosphate liberated can then be determined by the procedure described in the next section and the total phosphorus by the procedures of Ames *et al.*;[20] the amount of phosphonate phosphorus present is then obtained by subtracting the first result from the second.

I. Determination of Water-soluble Hydrolysis Products
of Complex Lipids

1. *Phosphorus*

The following method for the determination of phosphorus involves perchloric acid oxidation of phospholipids with the release of inorganic phosphate which is reacted with ammonium molybdate to form phosphomolybdic acid. This is reduced and the product determined spectrophotometrically.[121] The procedure described here is essentially that of Bartlett.[40]

"The ammonium molybdate reagent is prepared by dissolving the compound (4·4 g) in water (500 ml), adding concentrated sulphuric acid (14 ml) and making up to 1 l. The reducing reagent consists of sodium bisulphite (2·5 g), sodium sulphite (0·5 g) and 1-amino-2-naphthol-4-sulphonic acid (0·042 g) which are dissolved in water (250 ml) and allowed to stand in the dark for several hours. The solution is filtered into a brown bottle and, when refrigerated, is stable for about one month. A standard solution of 0·5 mм sodium dihydrogen phosphate is also prepared for calibration purposes.

Perchloric acid (0·4 ml; at least 70 per cent) is added to the dry lipid sample (containing up to 1 μmole of phosphorus) in a test-tube and the lipid is digested by gentle refluxing for 20 min on a heating block or sand-bath at a temperature such that the tube acts as an air condenser. The mixture is cooled and ammonium molybdate reagent (2·4 ml) is added followed by the reducing reagent (2·4 ml). The solutions are mixed thoroughly and are heated on a boiling water bath for 10 min for colour development. On cooling, the absorbance of the solution is measured at 820 nm. A blank sample is analysed simultaneously. The amount of phosphorus in the unknown sample is read from a calibration curve prepared at the same time by performing the reaction on known amounts of the standard phosphate solution."

It is possible to carry out the reaction in the presence of TLC adsorbents provided that the solution is centrifuged at approximately 800 × *g* and decanted for the measurement after colour development. Unacceptably high blank values may be obtained, however, unless the adsorbent is washed with solvent before the plates are made up (see Chapter 3).

When large numbers of samples must be analysed, it is possible to automate many of the steps.[121]

It should be noted that there is a slight danger of explosions occurring with perchloric acid digestions, especially if all traces of solvent are not removed. A safety screen should always be used.

2. *Glycerol*

A gas chromatographic procedure for the determination of glycerol in simple lipids has already been described (Chapter 6). The following method which involves acetolysis with acetic anhydride and trifluoro-acetic acid, saponification of the resulting diglyceride acetates and preparation of triacetin for GLC analysis is more suited to the determination of glycerol in complex lipids.[209]

"The phospholipid (up to 6 mg) is refluxed with acetic anhydride–trifluoroacetic acid (4:1, 2 ml) for 10 hr in the presence of a known amount of pentadecanyl acetate, which serves as an internal standard. Reagents are then removed *in vacuo* and 0·3 M sodium methoxide in methanol (2 ml) added and the mixture refluxed for 10 min. The methanol is removed by evaporation when water (0·3 ml) and 1 N sodium hydroxide (0·2 ml) are added and the mixture refluxed for 2·5 hr to complete the hydrolysis. Acetic anhydride (3·5 ml) and xylene (3·5 ml) are then added and triacetin is prepared as described earlier (Chapter 6)."

The conditions for gas chromatographic analysis of the product were also described in Chapter 6. Preliminary details of an improved procedure for glycerol analysis have been reported by Privett *et al.*[408]

3. *Carbohydrates, N-acetylneuraminic acid and related compounds*

(i) *Gas chromatographic methods*

The high sensitivity and resolving power of GLC makes it an ideal technique for the separation, identification and estimation of the carbohydrate moieties bound to complex lipids despite the fact that the acidic conditions necessary for the hydrolysis of glycosidic linkages result in the formation of all the anomeric forms of the methyl glycosides rather than chemically pure sugars. After conversion to the trimethylsilyl ether (TMS) derivatives, the various anomeric forms derived from galactose can be clearly separated by GLC from those derived from glucose and the amounts of these compounds are determined by summing the relevant peaks. The procedure now widely accepted was originally devised by Sweeley and Walker[532] and has recently been reviewed by Sweeley and Vance.[531] The glycosidic bonds are hydrolysed by reaction with anhydrous methanolic hydrogen chloride, methyl esters of the fatty acids formed are removed by extraction and, after removal of excess acid and the solvent, the TMS ether derivatives of the methyl glycosides are prepared. The following procedure is that of Sweeley et al.[531, 532] with a minor modification.[584]

"The glycolipid (up to 5 mg) in 0·75 N methanolic hydrogen chloride (2 ml) is heated in a sealed tube at 80°C for 24 hr. On cooling, hexane (2 ml) is added and, after thorough mixing and allowing to settle, the upper layer containing the methyl esters is discarded. Powdered silver acetate (0·25 g) is added and, after 2–3 min, the solution is filtered through a fine porosity filter, washing the filtrate with three 1 ml portions of methanol. The solvent is removed in a stream of nitrogen and the residual methylglycosides are converted to the TMS derivatives by the procedure described earlier (see Chapter 4; a reaction time of up to 15 min may be necessary, however)."

Mannitol may be added if necessary as an internal standard.

The method gives satisfactory results when galactose, glucose and N-acetylneuraminic acid are present in the lipid. On acidic hydrolysis, the last-named compound is converted to the 2-O-methyl ketal derivative of methylneuraminate which, on trimethylsilylation, appears as a single peak on the gas chromatograph. When N-acetylgalactosamine is present, difficulties may be encountered as this is partially hydrolysed to galactosamine, the TMS ether derivatives of the anomeric forms of which are eluted with those of galatose on GLC analysis. Before the TMS ether derivatives are prepared in this instance, it is therefore necessary to reconvert the galactosamine to the N-acetyl derivative by reaction with acetic anhydride in methanol by the method described above for the N-acetylation of long-chain bases (section 7F).

A GLC trace of the carbohydrate derivatives prepared by the above method from a ganglioside is illustrated in Fig. 7.10. SE-30 (2·5 per cent) serves as the stationary phase at a column temperature of about 160°C. Under these conditions, the three anomeric forms derived from galactose are eluted before the two anomeric derivatives from glucose. The principal peak from the N-acetyl galactosamine derivative and that of the neuraminic acid derivative have retention times of 2·4 and 4·2 respectively relative to the main glucose (the methyl α-glucoside derivative) peak. The TMS ether derivative of mannitol has a retention time of 1·5 relative to the methyl α-glucoside.

The response of the flame ionisation detector to the combined anomeric forms of all these compounds appears to be sufficiently similar for there to be no need for correction factors. As the response to the TMS derivative of mannitol is higher than that for the glucose derivative,

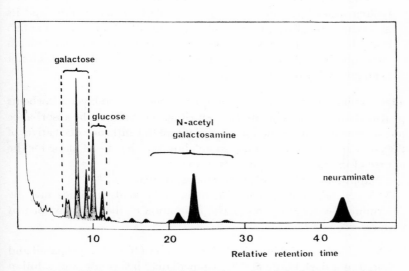

FIG. 7.10. GLC of trimethylsilyl derivatives of carbohydrates derived from glycolipids by acidic hydrolysis. Stationary phase—$2\frac{1}{2}$ per cent SE-30 at 160°C.[532] (Reproduced by kind permission of Dr. C. C. Sweeley and *Analytical Chemistry*.)

however, it is necessary to use a correction factor of 1·25 to compensate for this when mannitol is added as an internal standard.

Inositol in complex lipids can be determined by related procedures.[438, 576]

(ii) *Chemical procedures*

The amount of hexose units in glycolipids may be determined by converting the hexoses to the furfural derivatives in strong acid and reacting them with anthrone with which they form coloured complexes that can be estimated spectrophotometrically. The following method is essentially that of Yamamoto and Rouser.[606]

"A stock solution of 2 per cent anthrone in 98 per cent sulphuric acid (10 ml), which is stable for 3 weeks if refrigerated, is diluted with 87·5 per cent sulphuric acid (90 ml). This reagent (4 ml) is added to the glycolipid in dimethylformamide (1 ml) and the sample is heated in a boiling water bath for 4 min. The sample is cooled rapidly and its absorbance together with that of a reagent blank is measured at 625 nm. The amount of hexose present is determined from a calibration curve obtained ideally with a pure glycolipid standard or, if this is not available, with pure glucose or galactose."

The reaction can be carried out in the presence of thin layer adsorbents provided they are precipitated by centrifugation prior to the absorbance measurement. The extinction coefficients of the anthrone derivatives of the common sugars are similar so no allowance has to be made for the nature of the hexose components in the lipid.

α-Naphthol can be used in an analogous reaction.[606]

N-Acetyl neuraminic acid and other sialic acids in lipids may be determined by spectrophotometric estimation of the complex formed with resorcinol.[528]

"A stock solution of resorcinol (2 g) in water (100 ml) is prepared and stored in a dark bottle at 4°C. Immediately before use, this solution (10 ml), concentrated hydrochloric acid (80 ml) and 0·1 M copper sulphate solution (0·25 ml) are mixed and made up to a volume of 100 ml with water.

Samples containing up to $1\,\mu$mole of N-acetyl neuraminic acid are dried under vacuum (*iso*-amyl alcohol may be added to prevent foaming). Water (1 ml) and the resorcinol reagent (1 ml) are added and the mixture heated on a boiling water bath for 15 min and then chilled in ice. n-Butyl acetate and n-butanol (85:15, v/v; 2 ml) are added and, after thorough mixing and centrifugation, the absorbance of the upper layer of the sample and an appropriate reagent blank at 580 nm are determined. The amount of N-acetylneuraminic acid in the sample is calculated from a calibration curve prepared from the pure compound."

The procedure can also be applied to gangliosides containing N-glycolyneuraminic acid which has a greater molar extinction coefficient, but as a suitable standard may not be available for calibration purposes, values obtained from the N-acetylneuraminic acid calibration curve must be corrected by multiplying by a factor of 0·77. In an alternative procedure, gangliosides in natural lipid mixtures can be determined without prior isolation by a sensitive thiobarbituric acid assay of the sialic acid residues.[118]

Hexosamine,[527] inositol[14] and sulphate[121, 513] can also be determined by chemical procedures.

4. *Nitrogenous bases of phospholipids*

Five bases commonly occur in phospholipids—serine, ethanolamine, N-methylethanolamine, NN-dimethylethanolamine and choline. Chromatographic procedures have been developed for their analysis as they are metabolically related. They are eluted in the above order from columns of Dowex-50 ion exchange resin with 1·5 N hydrochloric acid.[108] This and other chromatographic methods for the analysis of these compounds have been reviewed by McKibbin.[325] Specific chemical methods have also been developed for the analysis of choline[574] and serine and ethanolamine.[119]

CHAPTER 8

The Analysis of Molecular Species of Lipids

A. Introduction

In nature, lipid classes do not exist as single pure compounds but rather consist of complex mixtures of related components in which the nature of the aliphatic residues varies from molecule to molecule. In some lipids, such as cholesteryl esters, only the single fatty acid moiety will vary; in others, for example triglycerides, each position of each molecule may contain a different fatty acid. Sphingolipids contain a number of different long-chain bases which may be linked selectively by amide bonds to specific fatty acids. A complete structural analysis of a lipid therefore requires that it be separated into *molecular species* that contain single specific alkyl moieties (fatty acids, alcohols, ether-linked aliphatic chains, long-chain bases, etc.) in all the relevant portions of the molecule. With lipids that contain only one or two alkyl groups, this is now often technically feasible. With lipids that contain more than two alkyl groups, means have yet to be developed for physically separating all the possible species that may exist although, if stereospecific enzymatic hydrolyses are performed on fractions separated by the available methods, it may be possible to at least calculate the amounts of all the molecular species that are present. The analyst must at the moment content himself with isolating simpler rather than single molecular species in such instances. Alkyl-, alkenyl- and acyl- forms of a given lipid are *not* molecular species of this and can themselves be fractionated into molecular species so should be isolated separately before an analysis of this kind is begun.

Ideally, it would be preferable if lipids could be separated into individual molecular species without being modified in any way so that the biosynthesis or metabolism of each part of the molecule could be studied with isotopically-labelled components. The technical problems of the analysis can often be considerably reduced, however, if the polar parts of complex lipids are rendered non-polar by the formation of suitable derivatives or are removed entirely by chemical or enzymatic means. Whichever approach is adopted, it is always necessary to apply combinations of different chromatographic procedures to achieve effective separations. These methods have been reviewed elsewhere.[289, 430, 560]

It should be remembered that it is almost always advisable to calculate the molar rather than weight proportions of any molecular species isolated. Precautions must always be taken to minimise the effects of autoxidation (see Chapter 3).

B. General Methods of Analysis

1. *Liquid–solid and liquid–liquid chromatography*

The principles of these procedures have been discussed in Chapter 3 and applications to the analysis of simple aliphatic molecules (fatty acids, alcohols, etc.) have been described in subsequent chapters so very little amplification is necessary here. It must be noted, however, that when such methods are applied to the isolation of molecular species of more complicated lipids, the separations achieved depend on the combined physical properties of all the aliphatic residues. If triglycerides are considered for illustrative purposes, adsorption chromatography will permit the separation of molecules containing three normal fatty acids from those containing two normal fatty acids and one fatty acid with a polar functional group in the chain from those containing one normal fatty acid and two polar fatty acids and so forth. Silver nitrate chromatography will separate those molecules containing three saturated fatty acids from those with two saturated fatty acids and one monoenoic acid which are in turn separable from a number of other distinct fractions containing molecules with fatty acids of a progressively higher degree of unsaturation. Liquid–liquid chromatography can be used to separate triglycerides in which the combined chain lengths of the fatty acid components differ by two carbon atoms (a double bond reduces the effective chain length of a fatty acid by two carbon atoms). The criteria that govern the separation of intact lipids are discussed in greater detail below.

Thin layer systems of liquid–solid and liquid–liquid chromatography are generally preferred over column methods for the separation of molecular species of lipids because of the greater resolving power of the former. A variety of procedures have been applied to the determination of fractions separated by this means. Charring techniques have been little used as fractions separated by the above methods are likely to differ considerably in the nature of their fatty acid constituents so that the amount of carbon formed is not necessarily linearly related to the amount of material present. Gas chromatography of the methyl esters of fatty acid constituents with an added internal standard is most often the method of choice as fractions are generally similar in polarity and, after transesterification, can be identified and estimated in a single analysis. This method may be combined with gas chromatography of the intact lipid, again in the presence of an appropriate internal standard, so that a check is obtained on the results. In addition as a check, the fatty acid composition of the unfractionated lipid can be computed by reconstituting the results from the individual fractions and this should agree with the original analysis.

2. *Gas–liquid chromatography of intact lipids*

(i) *Instruments and columns*

The principles of gas chromatography and applications of the method to comparatively simple molecules have been discussed in earlier chapters. Because of the low volatility of intact lipids, gas chromatographic analysis is beset with a number of difficulties and indeed for some time it was thought that molecules such as triglycerides with molecular weights up to 900 would pyrolyse at the temperatures required to elute them from GLC columns. Work from the laboratories of Kuksis and of Litchfield principally has shown that this need not occur and useful separations of such compounds can now be achieved. The procedure has been the subject of several reviews.[284, 287, 306, 308]

The gas chromatograph required is similar to that described earlier for fatty acid analysis but it is essential that it have a flame ionisation detector for maximum sensitivity and facilities for high precision temperature programming up to a temperature of at least 350°C. In addition, it should be of a construction such that on-column injection is possible, although a preheater is also necessary to warm up the carrier gas before it reaches the column packing. The dead volume between the end of the packing and the detector flame should be as small as possible and the flame jet should preferably be wider than normal so that comparatively high carrier gas flow rates can be used. Automatic flow controllers for the carrier gas are a useful accessory as the flow rate in short columns can change markedly on temperature programming. As bleeding of even the most thermally stable liquid phases occurs at high temperatures, better results are obtained with dual compensating columns than with single column instruments. The analyst should not be discouraged from using equipment that does not meet all these criteria, especially for separating compounds of intermediate molecular weight such as diglycerides or their derivatives, as patience and skill can compensate for many instrumental deficiencies.

Helium or nitrogen may be used as the carrier gas and excellent recoveries of high molecular weight triglycerides have been obtained with both, although the former is to be preferred whenever feasible as better resolutions are attainable at high flow rates.[308] Nitrogen is used most often for reasons of cost, but it is essential that it be of very high purity as traces of oxygen or water will destroy liquid phases at elevated temperatures.

Better results are obtained with glass columns than with those of other materials, but there may be technical difficulties in obtaining effective seals at each end of the column; O-rings of Viton or other materials generally become brittle and crack if used for any length of time at temperatures above 250°C. Fortunately some manufacturers are tackling the problem and efficient seals are now available for some makes of gas chromatographs. It is also possible to avoid this difficulty by using columns with direct glass to metal seals and these are also coming on to the market for specific instruments at moderate prices. Stainless steel columns can also be used if they are first silanised as the efficiencies obtained are then almost the same as those with glass columns.[290] Narrow bore columns give the best resolutions and those of $\frac{1}{8}$ inch diameter are preferable to those of $\frac{1}{4}$ inch; the length of the column selected will vary with the nature of the sample but it is usually necessary to compromise resolution by using short columns (50–100 cm) in order that compounds are eluted in a reasonable time.

The most useful liquid phases are silicone elastomers of high thermal stability such as SE-30, JXR and OV-1; the more polar silicone, QF-1, is occasionally employed and the newly introduced polymer, Dexsil 300, may offer some advantages. Kuksis[285, 288] has also obtained some promising results in separating diglyceride derivatives on polyester phases. Silanised solid supports (80–100 or 100–120 mesh) are essential and they are generally coated with low levels (1–3 per cent) of stationary phase by the filtration technique described earlier (Chapter 3). Care is necessary in preparing the columns which must be firmly packed to ensure adequate resolution, but not too tightly otherwise higher temperatures than are advisable may be necessary to elute samples in a reasonable time and losses of higher molecular weight components may occur. Finally, the column is sealed with a plug of silanised glass wool and conditioned at a temperature at least 25°C higher than that at which it is to be used for about 4 hr. It may be necessary to attempt the preparation of suitable columns several times before success is achieved and one should not be discouraged by initial failures.

(ii) *Operating conditions and nature of the separations*

The precise operating conditions and the resolutions attainable will vary with the nature of the samples to be analysed and they are discussed in the appropriate sections below, but some general comments can be made. With high molecular weight compounds such as triglycerides, short columns are essential and the column temperature must be programmed in the range 250–350°C at 2–5°/min. Slow programming rates give improved resolution generally. The optimum carrier gas flow rate will vary with the geometry of the instrument, the dimensions of the column and the amount of stationary phase on the packing material but will generally be 100 ml/min or greater. The sample, in solutions, should be injected by means of a syringe directly on to the column packing at a temperature about 40°C below that at which the first component emerges from the column. In this way, all the sample is vaporised but remains as a narrow band at the top of the column until the temperature programming procedure is under way. Samples containing about 20 μg of the largest component provide the optimum load. The entire analysis should be completed in 25–45 min. With samples of intermediate molecular weight such as diglyceride acetates, wax esters or cholesteryl esters, longer columns can be used to improve the resolutions attainable and the temperature limits of the analysis will be lower than with triglycerides, for example.

The silicone stationary phases which must be used for GLC analysis of high molecular weight compounds do not in general permit the separation of saturated from unsaturated components of the same chain length. Separations are then based solely on the approximate molecular weights of the compounds (for example, tripalmitin and myristopalmito-olein elute together) and components that differ by two carbon atoms in the combined chain lengths of the alkyl moieties must be separable before the columns are considered satisfactory. This can usually be achieved with column efficiencies of 500–1000 theoretical plates per foot of packing material and indeed, with well-packed columns, components differing in molecular weight by one carbon atom can often be completely resolved. A shorthand nomenclature is in common use to designate simply glycerides separated in this way; the total number of carbon atoms in the aliphatic chains of the compounds (but not in the glycerol moiety) are calculated and this figure is used to denote the compound. As an example, tristearin, triolein and trilinolein are referred to as C_{54} triglycerides or as having a *carbon number* of 54. Distearin has a carbon number of 36 and the diglyceride acetate prepared from it has a carbon number of 38 (although the two carbon atoms of the acetate moiety are not counted by some authors in calculating carbon numbers).

During isothermal operation, as discussed in Chapter 3, there is a logarithmic relationship between the retention time and carbon numbers of components of a homologous series, but when linear temperature programming is used there is a linear relationship between these parameters for short series of homologues. As a result, the elution temperature rather than the elution time is often quoted as a parameter that describes the retention characteristics of a given compound. With longer homologous series, the relationship begins to break down and improved resolutions are obtained with non-linear (concave) temperature programming profiles,[284] but as few commercial gas chromatographs are equipped with this facility, the refinement has not been widely adopted.

When flame ionisation detectors are used, the detector response is directly proportional to the weight of material eluting from the columns (see Chapter 3) and the amount of each component can be calculated from the areas of the peaks on the GLC recorder trace. There is no simple relationship between area, retention time and peak height for temperature programmed analyses so it is necessary to measure the area of each peak manually by one of the methods described earlier or, better, by means of an electronic digital integrator. Because of the high temperatures necessary for gas chromatography of intact lipids, there is always a danger that losses will occur on the columns as a result of pyrolysis or of reaction with the column materials. It is, therefore, necessary to check that acceptable reproducible recoveries are obtained and, if necessary, to calibrate the columns to compensate for any losses. It is not easy to check absolute recoveries from columns as this requires preparative facilities or means of counting radioactive samples eluting from the columns, but it is possible to check that recoveries are linearly related to the amount of material injected by inserting known quantities of standards (say 1–20 μg) into the columns and measuring the detector response. Alternatively, it can be assumed that in all but the very worst of columns, recoveries of tricaprin or trilaurin will be essentially complete so that standard mixtures of these triglycerides and higher molecular weight compounds can be analysed and relative losses determined. The losses that can be accepted will vary with the degree of difficulty of the analysis, but as a rough guide recoveries of the highest molecular weight component of a mixture of triglycerides, for example, should be at least 90 per cent relative to tricaprin. When pyrolysis occurs on the column, peaks for pure standards are often preceeded by broad humps of decomposed material.

As a wide range of unsaturated compounds is unlikely to be available for standardisation purposes, there are a number of advantages to be gained by hydrogenating all samples prior to analysis. If this is done, there are no selective losses of unsaturated components relative to saturated, peaks on the recorder trace are sharper, resolutions are improved and quantification of components is simplified. In the analysis of molecular species of lipids, it is necessary to know the molar proportions of all components separated and the weight responses of the detector must be corrected by multiplying by appropriate arithmetic factors obtained from the molecular weights of the compounds as described earlier for methyl esters of fatty acids (Chapter 5). Hydrogenation prior to analysis simplifies the range of factors required considerably and removes any dubiety about their numerical values. This is of course particularly important with samples containing polyunsaturated fatty acids.

Samples may be injected on to columns in carbon disulphide, diethyl ether, hexane or xylene solution; chloroform is also used on occasion but tends to strip the stationary phases from the packing and damage the flame ionisation detector. Hydrogenated lipids tend to be less soluble than the unsaturated compounds in the above solvents but will usually dissolve on warming.

(iii) *Preparative gas chromatography*

As the lipid fractions separated by the gas chromatographic techniques described above may contain a wide variety of fatty acid components, it would be very useful if compounds separated could be collected and further fractionated by other procedures or so that their fatty acid compositions could be determined. Kuksis and Ludwig[291] succeeded in doing this by adapting conventional preparative GLC techniques (see Chapter 5). Columns (2 ft × ¼ inch) containing higher amounts of stationary phase were used and it was necessary to inject the same sample a number of times and pool corresponding fractions to obtain sufficient material for further analysis. It was also necessary to rechromatograph fractions to obtain material of satisfactory purity. Although there were selective losses of unsaturated components, the fractions recovered were apparently not degraded in any way. The procedure is tedious and unreliable and has not been widely used.

C. Molecular Species of Simple Lipids

1. *Cholesteryl esters*

Cholesteryl esters are readily separated according to the degree of unsaturation of the single fatty acid moiety by TLC on adsorbents impregnated with silver nitrate. Separations are performed with the same TLC systems as were described earlier (Chapter 5) for methyl esters of fatty acids and fractions that contain fatty acids with zero to six double bonds are obtained in an analogous manner. Some separation of a similar nature is obtained on normal silica gel G plates with carbon tetrachloride or hexane–benzene (9:1, v/v) as solvents for the development.

Reverse phase TLC systems can also be used to separate cholesteryl esters containing critical pairs of fatty acids. For example, Kaufmann *et al.*[266] used silica gel G impregnated with liquid paraffin (see Chapter 3 for practical details) as stationary phase with methylethylketone–acetonitrile (7:3, v/v), 80 per cent saturated with liquid paraffin, as mobile phase and obtained resolutions comparable to those described earlier for methyl esters of fatty acids (Chapter 5).

In addition, cholesteryl esters can also be separated by high temperature GLC on columns similar to those used for triglyceride analysis (18 inches × $\frac{1}{8}$ inch with 2·25 per cent SE-30 as stationary phase) from which they elute at temperatures intermediate between those required for diglyceride acetates and triglycerides (see below). With temperature programmed analyses, compounds differing in molecular weight by one carbon atom are separable and with isothermal analyses, some separation of unsaturated components is possible (e.g. cholesteryl arachidonate and cholesteryl arachidate may be resolved).[283]

Single pure components that may be required for radioactivity measurements, for example, can be isolated preparatively by combining any two of the above procedures although the proportions of the various molecular species are obtainable simply from a fatty acid analysis alone.

2. *Monoglycerides*

1-Monoglycerides can be separated from 2-monoglycerides without isomerisation by chromatography on adsorbents impregnated with boric acid (see Chapter 6) and it is generally advisable to separate the two forms before proceeding to more involved analyses. It is customary to prepare volatile derivatives of monoglycerides such as acetates, trimethylsilyl (TMS) ethers or trifluoroacetates in addition to isopropylidene or benzylidene derivatives (see Chapter 4 for practical details), to reduce the polarity of the compounds and to prevent acyl migration. Trimethylsilyl ether derivatives of monoglycerides indeed are sufficiently volatile to be analysed by GLC on polyester columns such as EGS or PEGA which permit the separation of compounds differing in the chain lengths and degree of unsaturation of the fatty acid moieties.[593] 1- and 2-Monoglyceride derivatives are also separable on these columns and on non-polar stationary phases, the 2-acyl compound eluting first, but saturated and unsaturated compounds of the two types will overlap if an initial separation is not performed. Monoglyceride acetates can also be separated by molecular weight on columns of the type used to separate di- and triglycerides (see below) but at much lower temperatures.

Individual monoglyceride species could no doubt also be separated by silver nitrate or reverse phase TLC procedures. Mass spectrometry is a considerable aid in the identification of specific isomers.[246]

3. Diglycerides

Although diglycerides rarely occur in greater than trace amounts in tissues, they are important intermediates in the biosynthesis of other lipids. In addition, 1,2-diacyl-*sn*-glycerols are formed by the action of phospholipase C on glycerophosphatides (see Chapter 9) and, as these compounds or their derivatives are much more volatile and less polar than the parent complex lipids, it is common practice for molecular species of glycerophosphatides to be determined in this form. As a result, chromatographic methods for the analysis of diglycerides are of considerable significance.

Methods of separating α,β- and α,α'-diglycerides on boric acid impregnated adsorbents have been discussed earlier (Chapter 6) and it is advisable to carry out this separation before proceeding to more detailed analyses to simplify identification and improve resolution of components. Methods of fractionating α,β-diglycerides are discussed at length below but similar principles apply to the analysis of α,α'-compounds. They may be separated in the free state or as less polar stable derivatives such as the acetates or TMS ethers, but if derivatives are not prepared, it is advisable that the programme of analysis be performed as quickly as possible to minimise the risk of acyl migration. Acetates are the favourite derivatives as they are chemically stable on TLC adsorbents and during high temperature GLC, although the TMS ethers are also satisfactory for GLC purposes. A small amount of acyl migration occurs during the preparation of the derivatives none the less. Trifluoroacetate derivatives of diglycerides, however, decompose at the temperatures necessary to elute them from GLC columns.[285]

Diglyceride acetates are readily separated by silver nitrate TLC into simpler molecular species differentiated according to the degree of unsaturation of the combined fatty acid moieties. They migrate in the order:

$$SS > SM > MM > SD > MD > DD > ST > MT > STe > MTe > DTe$$
$$> SP > SH$$

where S, M, D, T, Te, P and H denote saturated, mono-, di-, tri-, tetra-, penta- and hexa-enoic fatty acid residues respectively (they do not indicate the relative positions of the fatty acids on glycerol as positional isomers are eluted together).[292, 412] It is noteworthy that one linoleate moiety is more strongly retained than two monoenoic moieties and one linolenate moiety is more strongly retained than two linoleate moieties. Figure. 8.1 illustrates the separation of diglyceride acetates prepared from phosphatidyl choline of pig liver on silica gel G impregnated with 10 per cent silver nitrate; the solvent system in this instance was chloroform–methanol (99:1, v/v) but hexane–diethyl ether (85:15, v/v) also gives satisfactory results. (N.B. to ensure that the solvent system contains the stated amount of methanol, it is first necessary to free the chloroform of any alcohol added as stabiliser (see Chapter 3).) When the sample contains high proportions of polyunsaturated fatty acids, it may be necessary to repeat the fractionation with a solvent system with an increased concentration of the more polar component. Diglyceride acetates containing up to twelve double bonds in the combined fatty acid moieties have been separated on similar plates with chloroform–methanol–water (65:25:4, by vol.) as the solvent system.[434] It is also possible to separate with care molecular species containing positional isomers of polyunsaturated fatty acids from each other. For example, diglyceride acetates containing 18:3 (*n*-6) migrate ahead of those con-

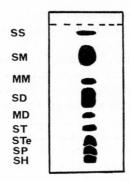

Fig. 8.1. TLC separation of diglyceride acetates prepared from the phosphatidyl choline of pig liver on layers of silica gel G impregnated with 10 per cent silver nitrate. Developing solvent: chloroform–methanol (99:1, v/v). Abbreviations: S, M, D, T, Te, P and H denote saturated, monoenoic, dienoic, trienoic, tetraenoic, pentaenoic and hexaenoic fatty acid residues respectively esterified to glycerol.

taining 18:3 (n-3) and species containing 20:4 (n-6) migrate ahead of those containing 20:4 (n-3) (Author, unpublished observations).

Approximately 10 mg of esters can be separated preparatively on a 20 × 20 cm TLC plate coated with a layer of the silver nitrate impregnated adsorbent 0·5 mm thick. Components are visualised under UV light after spraying with 2′,7′-dichlorofluorescein solution and may be identified and estimated by GLC after recovery from the adsorbent and transesterification with an added internal standard (see Chapters 3 and 6 for further discussion of the methods). Diglyceride acetates, free of dichlorofluorescein and silver compounds, can be recovered quantitatively from the adsorbent by the method of Akesson.[15]

"Bands are scraped from the plate into test-tubes and a solution of the internal standard, say methyl pentadecanoate, in methanol (1 ml) added to each followed by a 20 per cent aqueous solution of sodium chloride (1 ml) and hexane–diethyl ether (1:1, v/v; 3·5 ml). The contents are mixed thoroughly by means of a vortex mixer and centrifuged to precipitate the solids. The top layer is pipetted off and the bottom layer is washed twice more with similar volumes of hexane–ether. The combined extracts are washed with 0·05 M tris buffer (pH 9,

2 ml) and then with water–methanol (1:1, v/v; 2 ml) before being dried over anhydrous sodium sulphate."

After removal of the solvent, each sample is transesterified with sodium methoxide in the presence of dichloromethane or benzene which ensures that it remains in solution and reacts (see Chapter 4 for practical details). Methyl heptadecanoate may also be used as the internal standard. The molar amounts of the fractions are calculated by relating the combined areas of the fatty acid peaks on the GLC recorder traces, corrected from weight to molar amounts, to the area of the internal standard peak. In addition, tridecanoin may be added as the internal standard (in the hexane layer in this instance) and the intact compounds subjected to high temperature GLC as discussed below.

Trityl derivatives of diglycerides have also been separated by silver nitrate TLC when the amount of each fraction is determined simply by measuring the UV absorption of the trityl group.[126] Free diglycerides are separable by silver nitrate TLC on plates activated for longer periods than is usually deemed necessary (e.g. 4 hr at 120°C). With comparatively simple diglycerides prepared from the phosphatidyl ethanolamine of *E. coli*, it was possible to separate not only isomers differing in degree of unsaturation but homologues differing in chain length by two carbon atoms.[550] It is necessary to use slightly more polar solvent systems than with the acetate derivatives, e.g. chloroform–methanol (98:2, v/v) or hexane–diethyl ether (70:30, v/v).

Reverse phase systems have also been used to fractionate diglyceride acetates into molecular species containing critical pairs of fatty acid combinations; for example 16:0–16:0 migrates with 16:0–18:1, but both are separable from 16:0–18:0. Silicone (Dow Corning 200) liquid on silica gel G was the stationary phase and acetic acid–water mixtures the mobile phase[430] (a system originally devised by Privett and Blank).[407]

Although free diglycerides can be subjected to gas chromatography, some rearrangement and transesterification inevitably occurs so it is more usual for the acetate derivatives to be prepared. The nature of the columns and the precise operating conditions that have been recommended vary somewhat from laboratory to laboratory but as a rough guide 1 m × 3 mm o.d. columns packed with 2 per cent SE-30 on a silanised support will give excellent results. Carrier gas flows of 100–200 ml/min and temperature programming from 220–280°C are also commonly quoted. Some deviation from these optimum conditions is permissible, however, and the separations illustrated in Fig. 8.2 were obtained on 50 × 0·6 cm glass columns containing 1 per cent SE-30, temperature programmed from 250 to 300°C with a carrier gas flow of 50 ml/min of nitrogen. Components are identified by their retention times (or elution temperatures) relative to standard triglycerides having the same total number of carbon atoms. The resolutions obtained are sufficient to separate components with combinations of fatty acids differing in chain length by one carbon atom. Although the separations depend largely

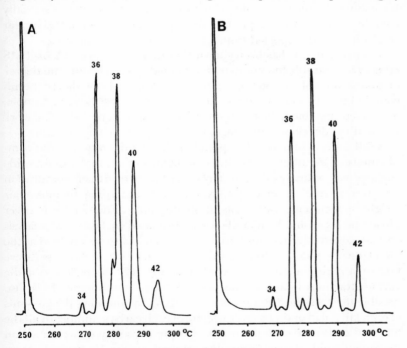

FIG. 8.2. GLC separation of diglyceride acetates prepared from the phosphatidyl choline of pig liver. 50 × 0·6 cm glass column packed with 1 per cent SE-30 on chromosorb W (acid-washed and silanised; 100–120 mesh); temperature programmed from 250–300°C at 2°/min; carrier gas—50 ml/min nitrogen. Trace A before and trace B after hydrogenation. The numbers above each peak refer to the "carbon number" of the component.

on the molecular weights of the components, some partial separation on the basis of degree of unsaturation may also occur but may not be especially desirable. For example, in Fig. 8.2(A), the components of carbon number 38 consists of two partially resolved peaks; the first contains 16:0 together with 20:4 and the second consists of two C_{18} acids of varying degrees of unsaturation. On hydrogenation (a suitable procedure is described in Chapter 5), these components merge and the remaining peaks are distinctly sharper with improved resolution (Fig. 8.2B); in particular, trace amounts of components containing odd-chain fatty acids become apparent. With these particular columns and operating conditions, recoveries of a wide range of standards were essentially complete whether the compounds were hydrogenated or not and similar results have been reported from a number of other laboratories.

More recently, it has been shown that diglyceride acetates or TMS ethers are separable to some extent both by the chain length and degree of unsaturation of the combined fatty acid constituents on short columns containing polyesters as liquid phases;[285, 288] EGSS-X for example, is sufficiently stable to be taken up to 300°C for brief periods. The work is still at the developmental stage, but holds promise for the future.

A full analysis of molecular species of diglycerides requires that combinations of the above procedures be used. For example, diglyceride acetates may be analysed separately according to degree of unsaturation by silver nitrate chromatography and then according to molecular weight by high-temperature gas chromatography. Fractions from silver nitrate plates should be identified and estimated by fatty acid analysis with an added internal standard but then should also be analysed as the intact compounds (hydrogenated if necessary) by high-temperature GLC with tridecanoin as internal standard. The proportions of the various fractions found by the two methods should agree. It is also possible to check the accuracy of the analyses by summing the amounts of each fatty acid or diglyceride acetate of each carbon number in all the fractions and comparing the results with similar analyses performed on the intact lipid. *Molar* proportions should always be quoted. Mass spectrometry has recently been applied to the identification of diglyceride derivatives eluted from GLC columns.[228] As an alternative to GLC, reverse phase chromatography has certain advantages especially in preparative applications.

The main limitation of all these procedures is that components with two given fatty acids in different positions of the glycerol moiety (e.g. 1-palmito-2-olein and 1-oleo-2-palmitin) are not separated. The amounts of various isomers of this type can, however, be calculated if each fraction is subjected to enzymatic hydrolysis with pancreatic lipase (see Chapter 9).

Alkyl- and alkenyl-analogues of diglyceride acetates have also been prepared and analysed by similar techniques to those described above.[428, 429]

4. *Triglycerides*

As triglycerides are the most abundant lipid class in nature and all the commercially important fats and oils consist almost entirely of this lipid, more effort has been applied to the determination of molecular species of triglycerides than to any other single lipid class. As each molecule may contain three different fatty acids, the problem can be extremely complicated; for example, a triglyceride with only five different fatty acid constituents may consist of 75 different molecular species (not including enantiomers). Several review articles have appeared recently[328, 392, 560] and Litchfield[306] has written the definitive work on the analysis of triglycerides.

Triglycerides, particularly those of seed oils, may contain a wide range of fatty acids including those of very short chain length (from acetic acid upwards) or those containing polar functional groups. In such instances, useful separations of molecular species of triglycerides can often be obtained by adsorption chromatography. For example, triglycerides of the Beluga whale contain isovaleric acid and species containing zero, one and two molecules of this acid have been isolated by TLC on silica gel G with a developing solvent of hexane–diethyl ether–acetic acid (87:12:1, by vol.);[307] the presence of a short-chain acid retards the migration of the molecule (Fig. 8.3, plate A). Similarly, the triglycerides of ruminant milks can be separated by related TLC systems into two fractions—one containing three fatty acids of normal chain length and one containing two normal fatty acids and one short-chain fatty acid. Polar fatty acid constituents also retard the migration of triglycerides on adsorbents and triglycerides from many seed oils that contain epoxy

I

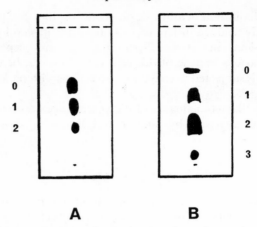

A   **B**

FIG. 8.3. TLC separation of triglycerides containing short-chain or polar fatty acids on layers of silica gel G. Plate A, Beluga whale oil triglycerides—the numbers refer to species with 0, 1 and 2 iso-valeroyl residues per mole. Solvent system: hexane–diethyl ether–acetic acid (87:12:1 by vol.).[307] Plate B, *Euphorbia lagascae* seed oil—the numbers refer to species with 0, 1, 2 and 3 epoxy-fatty acid residues per mole. Solvent system: hexane–diethyl ether (3:1, v/v).

fatty acids, for example, can be separated into simpler species that contain zero, one, two and three epoxy-fatty acid residues per mole of glycerol.[104] A solvent system of hexane–diethyl ether (3:1, v/v) with silica gel G layers, for example, was used to isolate such fractions from the seed oil of *Euphorbia lagascae* as illustrated in Fig. 8.3, plate B. Similar separations have been obtained of triglycerides that contain fatty acids with hydroxyl or other polar functional groups. Estolides can also be separated in this way from normal triglycerides.[357]

GLC of the fatty acid constituents in the presence of a suitable internal standard (e.g. methyl pentadecanoate or methyl heptadecanoate) would normally be the method of choice for estimating the amounts of components isolated by adsorption procedures. The amounts that can be separated preparatively in this way vary with the nature of the fatty acid constituents but generally, the more polar the functional groups in the fatty acid, the more that can be applied to a TLC plate. As a rough guide, 10–25 mg of material can be applied to a 20×20 cm plate coated with a layer 0·5 mm thick of silica gel G.

Countercurrent distribution has also been used to separate triglyceride species containing polar fatty acids, but the same fractions can usually be obtained with very much less effort, though admittedly on a smaller scale, by adsorption chromatography.

Silver nitrate chromatography has proved immensely useful in separating triglycerides containing a more normal range of fatty acids with zero to three *cis* double bonds into simpler molecular species. A triglyceride of this type can contain species with up to nine double bonds in the fatty acid moieties per mole of glycerol. Components migrate in the order:

SSS > SSM > SMM > SSD > MMM > SMD > MMD >
SDD > SST > MDD > SMT > MMT > DDD > SDT >
MDT > DDT > STT > MTT > DTT > TTT

where S, M, D and T denote saturated, mono-, di- and tri-enoic acids respectively[171, 441] (they do not indicate the positions of the fatty acids on the glycerol moiety) although there may be some changes in this order depending on the nature of the solvent mixtures used for development. Some simplification of even more highly unsaturated triglycerides such as fish oils has also been attained by silver nitrate TLC.[56] Silver nitrate (10–15 per cent) is incorporated into the layers and the solvent systems employed generally consist of hexane–diethyl ether, benzene–diethyl ether or chloroform (alcohol-free)–methanol mixtures. As all the fractions listed above cannot be separated on one plate, it is common practice to separate the least polar fractions first with hexane–diethyl ether (80:20, v/v) or chloroform–methanol (197:3, v/v) as illustrated in Fig. 8.4 (plate A) and then to separate the remaining fractions with more polar solvents such as diethyl ether alone or chloroform–methanol (96:4, v/v) as illustrated in Fig. 8.4 (plate B). Bands are detected under UV light after spraying with 2′,7′-dichlorofluorescein solution and components are recovered and are identified and determined by gas chromatography of the fatty acid constituents with an added internal standard as discussed earlier for diglyceride acetates. Approximately 10 mg of triglyceride can be separated on a 20×20 cm plate (0·5 mm thick layer) and excellent separations of large numbers of components have been obtained with 20×40 cm plates.

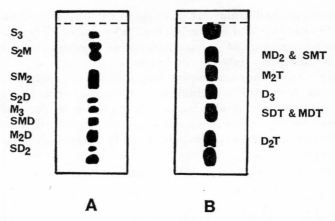

FIG. 8.4. TLC separation of soyabean triglycerides on layers of silica gel G impregnated with 10 per cent (w/w) silver nitrate. Developing solvent; Plate A, Chloroform–methanol (99:1, v/v). Plate B, Chloroform-methanol (96:4, v/v). Abbreviations: S, M, D and T denote saturated, monoenoic, dienoic and trienoic fatty acid residues respectively esterified to glycerol.

In addition, some separation of isomeric compounds is possible. For example, triglycerides of the type S_2M in which the monoenoic component is in position 2 are separable with care from the related compounds in which the monoenoic compound is in position 1 or 3.[39] Presumably, the presence of long-chain fatty acids on either side of the monoenoic acid weaken the π-complex with the silver ion permitting the component with the monoenoic fatty acid in position 2 to migrate ahead of isomers in which this residue is in position 1 as separations of this kind cannot be obtained with diglyceride acetates where one side of the molecule is comparatively open.

Column chromatography with silver nitrate impregnated adsorbents has also been used to separate triglycerides, but it lacks the resolving power of thin-layer chromatography. When the fatty acid constituents of triglycerides contain *trans*-double bonds, the elution pattern is much more complicated as components with *trans*-acids migrate ahead of analogous compounds with fatty acids containing *cis*-double bonds producing a complex elution pattern. Molecular species of triglycerides with polar fatty acid constituents, separated by adsorption chromatography as described above, may be further fractionated by silver nitrate TLC.[104]

High-temperature GLC can also be used to obtain efficient separations of triglycerides on a molecular weight basis although the conditions necessary to elute them from the columns approach the limits of thermal stability both of the stationary phases and of the compounds themselves. Although triglycerides with carbon numbers up to 68 have been successfully resolved,[192] there is little margin for error in the preparation of columns and in most laboratories, it is considered sufficient of an achievement to adequately resolve components with carbon numbers up to 56 or 58. Short columns (50 to 100 cm $\times$ 3 mm), high carrier gas flow rates and low loads of stationary phases (1–2 per cent SE-30 or JXR) are normally used with temperature programming at 2–5°/min from 180° to 350°C, depending on the range of molecular weights of the compounds to be analysed. Individual components are recognised by their carbon numbers relative to authentic standards. They are separable into homologues differing in carbon number by two units with relative ease but more care is necessary to separate components differing by one unit when odd-chain fatty acids are present in the sample.[309] Figure 8.5 illustrates the kind of separation obtainable with two natural triglyceride samples, A—coconut oil, B—pig adipose tissue triglycerides (lard), although the operating conditions in this instance were not those generally considered optimal. Efficient separations of the lower molecular weight components of the former sample are readily achieved but it is less easy to effectively separate the higher molecular weight components in the latter. Again, if the compounds are hydrogenated prior to the analysis, resolutions are improved and losses are reduced as saturated compounds are more stable to the operating conditions. Poor resolutions are inevitable with natural triglycerides such as ruminant milk fats that contain odd-chain and branched-chain fatty acids as well as fatty acids with a very wide range of chain lengths.

Quantification must be carefully checked with standard mixtures to ensure that losses of high molecular weight components relative to those of lower molecular weight are as low as possible; for example, if the recovery of a C_{54} standard is less than 90 per cent that of a C_{30} standard, the column is not satisfactory. Greater losses of higher molecular weight triglycerides such as trierucin (C_{66}) are inevitable, however, and must

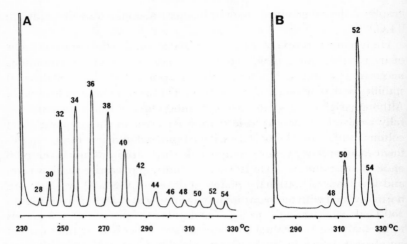

Fig. 8.5. GLC separation of intact triglycerides. A, Coconut oil triglycerides. B, Pig adipose tissue triglycerides. Column as in Fig. 8.2; temperature programmed from 230–330°C at 2°/min (A) and 280–330°C at 2°/min (B). The numbers above each peak refer to the "carbon number" of the component.

at present be accepted, but all such losses can be compensated for by determining calibration factors, provided that they are checked regularly. Results obtained from the GLC recorder traces should always be converted to molar proportions by multiplying by appropriate arithmetic factors.

GLC of triglycerides is used almost exclusively as an analytical technique although some preparative applications have been described[291] as discussed above.

A number of liquid–liquid partition systems have been devised for the separation of intact triglyceride molecules into simpler molecular species. Components, separated both by molecular weight and by the number of double bonds, are sometimes defined by a "partition number" where

partition number = carbon number $-2 \times$ (number of double bonds)[305]

Triglycerides having the same partition number, such as distearoolein and distearopalmitin, travel together in liquid–liquid systems. Reverse phase TLC has been used more frequently than column chromatography because of the greater resolution attainable with the former. For example, Litchfield[305] separated triglycerides differing in partition number over a range of 38 to 48 efficiently on 20×40 cm plates with silanised silicic acid impregnated with 8 per cent (w/w) hexadecane as stationary phase while nitroethane saturated with hexadecane was the mobile phase. 4–10 mg of Triglycerides were separated preparatively in this way and bands were detected by exposing the edge of the plate to iodine vapour; adjacent areas, not contaminated with iodine, were scraped from the plate for analysis (although there is a report that 2′,7′-dichlorofluorescein sprays can also be used in related circumstances).[581] A separation of this kind is illustrated in Fig. 8.6. Kaufmann *et al.*[265, 267] have described a number of similar reverse phase systems in which tetradecane on silica gel G is the stationary phase and acetone–acetonitrile (80:20, v/v), 80 per cent saturated with tetradecane, is the mobile phase. Litchfield's procedure may have particular advantages with highly unsaturated fats such as those of fish oils that are less easily analysed by silver nitrate TLC. Considerable practice is necessary with the technique, however, before separations such as that illustrated (Fig. 8.6) can be obtained routinely.

The column procedure of Nickell and Privett described earlier

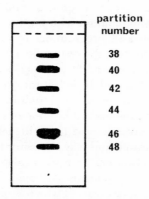

Fig. 8.6. Reverse phase TLC separation of *Ephedra nevadensis* triglycerides (fatty acids from 14:0 to 20:4 are present). Stationary phase—silanised silicic acid impregnated with 8 per cent hexadecane; mobile phase—nitroethane saturated with hexadecane.[305]

(Chapter 5) in which heptane on silanised celite is the stationary phase and acetonitrile–methanol (85:15, v/v) saturated with heptane is the mobile phase has been used to effect remarkable separations of triglycerides.[377, 385] By this means, components differing in partition number by one unit and species containing fatty acids with *trans*-double bonds, that could not be resolved by silver nitrate TLC, were separated. The procedure is tedious and does not appear to have been used other than in Privett's laboratory but could undoubtedly make a significant contribution to lipid analysis if adapted to modern liquid–liquid chromatography instruments.

A mass spectrometric system of determining molecular species of triglycerides has also been devised[206] in which mass spectra, obtained from a sample placed directly in the ion source of the mass spectrometer, are fed into a computer programmed to recognise characteristic ions, calculate the proportions of various molecular species (separated by the chain length and degree of unsaturation of the constituent fatty acids) from the ion intensities and print out the results. Large numbers of samples can be handled routinely, but few lipid analysts have such equipment available; the procedure is not suitable for triglycerides labelled with radio-isotopes.

In order to obtain meaningful separations of simpler molecular species of triglycerides, it is necessary to employ a combination of techniques. Silver nitrate TLC in combination with high temperature GLC is the most popular approach and programmes of analysis (including the various checks) similar to that discussed above for diglyceride acetates are commonly adopted. Litchfield[305] has described an alternative integrated system in which liquid–liquid chromatography is followed by high-temperature GLC and the former in combination with silver nitrate TLC is also used. Again, the main limitation of such procedures is that positional isomers of triglycerides with specific fatty acid combinations are not in general separable and it is necessary to apply stereospecific analysis procedures (see Chapter 9 for practical details) to each fraction to determine all possible components. Although this is now technically feasible,[186] the difficulties, not least of which is obtaining each fraction in sufficient quantity for analysis in a pure state, are enormous.

Molecular species of alkyl diglycerides and neutral plasmalogens have been isolated by related procedures to those described above.[590, 597]

5. *Wax esters*

In wax esters there are long aliphatic chains in both the fatty acid and alcohol moieties, each of which may contain double bonds, and silver nitrate TLC may be used to separate components according to the number of double bonds in total in both parts of the molecule. Solvent systems[1,175] similar to those described above for cholesteryl esters and methyl esters (Chapter 5) are used. Wax esters have similar molecular weights to diglyceride acetates and are eluted under comparable conditions from GLC columns.[1,237] In this instance they are easily separated into components differing in carbon number by one or two units, where the carbon atoms of the alcohol and fatty acid moieties have equal value. Mass spectrometry may be used to identify and estimate compounds eluting in a single peak (after deuterohydrazine reduction of any double bonds); characteristic ions at R. CO_2H^+ and R.$CO_2H_2^+$ (from the fatty acid residue) and at $(R' - 1)^+$ (from the alcohol moiety) are measured.[1] Reverse phase TLC (tetradecane as stationary phase and acetone–methanol, 9:1, v/v, as mobile phase) has also been used to separate critical pairs of wax esters, e.g. palmityl palmitate, palmityl oleate and oleyl palmitate migrate together.[268]

Again, an integrated programme of analysis using at least two of the above procedures (as discussed earlier for diglyceride acetates) should ideally always be applied to wax esters.

D. Molecular Species of Complex Lipids

1. *General approaches to the problem*

Molecular species of complex lipids such as glycerophosphatides are most easily separated after the phosphorus moiety has been removed by enzymatic hydrolysis with phospholipase C (see Chapter 9 for practical details); the resulting diglycerides (1,2-diacyl-*sn*-glycerols) are normally purified and acetylated *immediately* after preparation to prevent acyl migration during subsequent separation procedures. Indeed, even the diglyceride acetates should be purified by TLC on silica gel G (hexane–diethyl ether, 70:30, v/v is a suitable solvent system) before being analysed further. Diglyceride acetates from all phospholipids may be separated into molecular species by the procedures discussed earlier in this chapter. Alkyl- and alkenyl- forms should be isolated (see Chapter 7) before more detailed studies are commenced.

No satisfactory chemical method for removing the phosphorus moiety of glycerophosphatides has been developed, although acetylation with equal mixtures of acetic anhydride and acetic acid at 140°C in a sealed tube has been used for the purpose.[425] Unfortunately acyl migration occurs with formation of 1,3-diglycerides and the 1,2-diglyceride acetates obtained are not sufficiently similar in structure to the original phospholipid to warrant more detailed analyses.[382, 427] Similarly, trimethylsilyl-ether derivatives of diglycerides have been prepared from glycerophosphatides by subjecting the latter to very high temperatures (approx. 250°C) for brief periods and then silylating or by reacting them with the silylating reagents for prolonged periods at high temperatures.[228, 229] Here also, 1,3-derivatives accompany the 1,2-compounds and the latter do not accurately represent the original composition of the native phospholipid so that their use in structural analyses is limited.

In some studies it may be necessary to retain the phosphorus moiety in the molecule (e.g. in metabolic studies of compounds labelled with ^{32}P) and methods have been developed for isolating molecular species of certain glycerophosphatides in the native state; alternatively, the polar head group may be rendered less polar by conversion to appropriate derivatives and the glycerophosphatides separated into molecular species in this form. Silver nitrate TLC and reverse phase TLC, preferably used to complement each other, are then the methods of choice for isolating simpler species, but GLC cannot be used as glycerophosphatides and their derivatives apparently pyrolyse at the temperatures necessary to elute them from the columns. In general, resolutions obtained with compounds that still contain the phosphorus moiety are slightly inferior to those that can be achieved with diglyceride acetates. As was the case with simple lipids, it may be necessary to subject any fractions isolated to hydrolysis with stereospecific lipolytic enzymes in order to obtain a complete picture of the lipid structure.

Glycosyl diglycerides and sphingolipids can sometimes be separated into simpler molecular species by related pathways.

2. *Phosphatidyl choline*

Molecular species of phosphatidyl choline have been analysed via the diglyceride or diglyceride acetate derivatives in a number of laboratories.[292, 384, 428, 549] The phospholipase C of *Clostridium welchii* (see Chapter 9) is generally used in their preparation. In addition, alkyl- and alkenyl- forms of phosphatidyl cholines have also been separated in this way.[428]

There have been a number of attempts to separate species of native phosphatidyl choline on thin layers impregnated with silver nitrate and most have met with very little success, probably because the highly polar phosphorus-containing group masked the comparatively small changes in polarity produced by the formation of π-complexes between silver ions and the double bonds of the unsaturated fatty acids. Arvidson,[31, 33] however, has achieved some valuable separations by using highly active adsorbent layers; 0·35 mm thick layers were prepared with silica gel H and silver nitrate in the proportions 10:3 (w/w) and were air-dried at room temperature (in the dark) for 24 hr and then at 175°C for 5 hr or at 180°C for 24 hr. Plates prepared under the latter conditions were considerably more active than those dried at the lower temperature. Using a solvent system of chloroform–methanol–water (60:35:4, by vol.) and the more active plates, species with one to six double bonds per molecule were separable from each other. Only the saturated and monoenoic fractions were not completely resolved but they could be completely separated on the less active plates with the same solvent system. Up to 20 mg of phospholipid could be separated into molecular species on a 20×20 cm plate. Typical separations are illustrated in Fig. 8.7.

Lipids were recovered from the adsorbent after detection with a 2′,7′-dichlorofluorescein spray by elution with chloroform–methanol–acetic acid–water (50:39:1:10 by vol.) which was washed with one third of a volume of 4 M ammonia to remove dye and excess silver ions. Fractions were determined by phosphorus analysis.[32, 33]

The complete removal of silver ions occasionally presents some difficulty, but the above procedure is as effective as most that have been attempted. Rechromatography on silica gel H has been used for the purpose, but then losses inevitably result. Alternatively, passage through

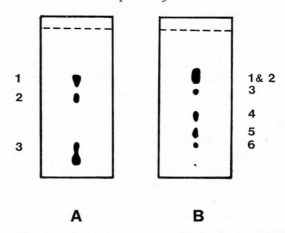

Fig. 8.7. Fractionation of native phosphatidyl cholines by silver nitrate TLC[31, 33] (layers of silica gel H–silver nitrate, 10:3, w/w). Solvent system, chloroform–methanol–water (65:25:4 by vol.). Plate A activated at 180°C for 24 hr; plate B activated at 175°C for 5 hr. The numbers alongside the plate refer to the number of double bonds in each species.

a cation exchanger (e.g. AG-50W-X8, 200–400 mesh, H$^+$ form; Biorad Laboratories, U.S.A.) in methanol–acetic acid–water (90:1:5 by vol.) is effective,[295] but fractions may be contaminated by organic materials leaching from the resin. It has also been reported that if the fractions are transesterified in the presence of the silica gel with methanol and sulphuric acid, silver ions are removed during the aqueous washing step;[207] the internal standard procedure and gas chromatography can then be used to determine the amounts of each fraction.

Arvidson[32] has also successfully separated native phosphatidyl cholines by reverse phase TLC into components of the same partition number. Undecane on silanised Kieselguhr G was the stationary phase and methanol–water (9:1, v/v), 70 per cent saturated with undecane, was the mobile phase. Each of the fractions from silver nitrate plates was separable into two components, largely those species that contained 16:0 and those that contained 18:0 (approximately 2 mg of material per 20 × 20 cm plate). Bands were visible without spraying as grey areas on a white background and could be recovered from the plate by elution with chloroform–methanol (2:1, v/v). "Green fingers" and practice are necessary before reproducible results can be obtained with either of the above two procedures.

Other workers have attacked the problem by rendering the head group less polar, i.e. phosphatidyl choline is converted enzymatically to phosphatidic acid by reaction with phospholipase D and this is in turn converted to the dimethyl ether by reaction with diazomethane (the preparation of these compounds is discussed in greater detail in Chapter 9). The use of such derivatives to isolate alkyl- and alkenyl- forms of phosphatidyl choline is discussed in Chapter 7. Dimethyl phosphatidates are separable by silver nitrate and reverse phase TLC (but not by high-temperature GLC) with almost the same facility as diglycerides. For example, Renkonen[433] used silver nitrate TLC to obtain distinct fractions with zero to six double bonds per molecule. Chloroform–methanol (98:2, v/v) separated the less unsaturated species and chloroform–methanol–water (90:10:1 by vol.) the more unsaturated species. Reverse phase TLC has also been used[601] to separate dimethylphosphatidates; tetradecane on Kieselguhr G was the stationary phase and acetonitrile–acetone–water (8:1:1 by vol.) saturated with tetradecane was the mobile phase.

3. *Phosphatidyl ethanolamine*

Phosphatidyl ethanolamine is most easily separated into molecular species in the form of diglycerides or diglyceride acetates, prepared by reaction with the phospholipase C of *Bacillus cereus* (see Chapter 9 for practical details) and this approach has been followed in a number of laboratories.[218, 426, 550, 597]

Analogous species of native phosphatidyl ethanolamine to those of phosphatidyl choline, can be isolated by the silver nitrate and reverse phase procedures of Arvidson,[33] described above, under exactly the same conditions. A similar approach was used by others.[226] It is also possible to convert phosphatidyl ethanolamine to dimethyl phosphatidic acid by the sequence of reactions detailed earlier and separate this into molecular species by the procedures used with the same derivatives prepared from phosphatidyl choline.[433] Non-polar N-dinitrophenyl-O-methyl derivatives of phosphatidyl ethanolamine (see Chapter 7 for details of preparation) are also separated into simpler species comparatively readily by silver nitrate TLC.[431]

4. *Other glycerophosphatides*

Molecular species of a number of other glycerophosphatides have been isolated by the diglyceride acetate approach including phosphatidyl inositol,[16, 220, 591] polyphosphoinositides,[220] phosphatidyl serine,[591] phosphatidyl glycerol,[193] cardiolipin[271b, 591] and phosphatidic acid.[202] Acidic or highly ionic glycerophosphatides are normally hydrolysed to diglycerides more readily by the phospholipase C of *B. cereus*, but phosphatidyl inositol and polyphosphoinositides have also been dephosphorylated by means of a specific phosphodiesterase from brain[220] and phosphatidic acid has been dephosphorylated with a phosphatidic acid phosphatase.[202] The phospholipase C of *B. cereus* releases both diglyceride moieties from cardiolipin. The enzymes used in these structural studies are discussed in greater detail in Chapter 9.

Many other glycerophosphatides may be separated into molecular species in the form of the dimethylphosphatidates. For example, phosphatidic acid [402] itself has been analysed in this form and phosphatidyl serine and phosphatidyl glycerol are both hydrolysed rapidly by phospholipase D so could undoubtedly be analysed in this way also. Although phosphatidyl inositol is not readily hydrolysed by phospholipase D, it can be converted to phosphatidic acid by reaction with periodate.[318]

Native phosphatidyl inositol has been fractionated by silver nitrate TLC on highly active plates with chloroform–methanol–water (65:35:5, by vol.) as solvent system[219] but mono- and dienoic species were not completely resolved. Luthra and Sheltawy,[318] however, were able to retain the phosphorus and inositol moieties and improve the resolutions attainable by another approach. The inositol residue was acetylated with acetic anhydride and pyridine yielding a triacetyl derivative as the main product. After purification, the phosphoric acid moiety was rendered non-polar by reaction with diazomethane when the resulting compound was separable into distinct species with zero to six double bonds per mole of phospholipid by silver nitrate TLC with chloroform–

acetone mixtures (75:25 or 50:50, v/v) as developing solvent. Two novel procedures were used to estimate the fractions separated. In the first, each fraction was transesterified in the presence of [³H] methanol and in the second, the lipid was acetylated with [1 − ¹⁴C] acetic anhydride; the [³H] methyl esters or [1 − ¹⁴C] acetyl phosphatidyl inositol derivatives were then determined by scintillation counting.

Attempts have been made to fractionate native phosphatidyl glycerol although with only limited success, but the procedures used for separations of species of phosphatidyl inositol could conceivably be of value with phosphatidyl glycerol also.

5. *Glycosyl diglycerides*

No enzyme has yet been isolated that hydrolyses the glycosidic linkage in glycosyl diglycerides and molecular species of these compounds cannot be determined via the diglyceride acetates. However, satisfactory procedures exist for separating the intact lipids into simpler species: for example, monogalactosyl diglycerides have been separated into distinct fractions containing one to six double bonds per molecule on thin layers of silica gel H containing 10 per cent (w/w) silver nitrate with chloroform–methanol–water (60:21:4, by vol.) as solvent system.[376] Somewhat better resolutions of both mono- and digalactosyl diglycerides are obtained if they are first acetylated with acetic anhydride and pyridine; the resulting compounds are similar in their chromatographic behaviour to diglyceride acetates and have been separated into seven clear fractions containing zero to six double bonds per mole on silica gel G plates impregnated with 5 per cent (w/w) silver nitrate with a developing solvent consisting of chloroform–methanol (197:3, v/v).[30] High-temperature GLC of the acetate derivatives does not appear to have been attempted, but trimethylsilyl ether derivatives of mono- and digalactosyl diglycerides (after hydrogenation) may be separated into species differing in the molecular weights of the component fatty acids by gas chromatographic conditions similar to those used with triglycerides.[35]

6. *Sphingomyelin and glycosphingolipids*

In Chapter 7, it was noted that sphingomyelin is readily separable into two species—one containing very long-chain fatty acids and one containing fatty acids of medium chain length. Glycosphingolipids can sometimes be partially separated into two classes—one containing normal fatty acids and one containing 2-hydroxy fatty acids (see Chapter 7), but molecular species of intact sphingolipids are not easily isolated otherwise. Considerable simplification of sphingolipid species containing various long-chain bases and fatty acids in combination can be achieved, however, if they are first converted to the less polar ceramides (see Chapter 7 for practical details). Ceramides of sphingomyelin have received most attention as satisfactory methods of preparing such compounds from all glycolipids have yet to be devised.

Karlsson and Pascher[257] have published the definitive paper on thin-layer chromatography of ceramides. There may be from two to four hydroxyl groups in the molecule (two or three in the base and zero or one in the fatty acid) and TLC on silica gel or better on silica gel containing diol complexing agents such as sodium tetraborate ($Na_2B_4O_7$. $10H_2O$) or sodium *meta*-arsenite ($NaAsO_2$) can be used to effect separations that depend on the number and configuration of the hydroxyl groups present. Layers are prepared by incorporating 1 per cent (w/v) of the salt into the water used to prepare the slurry of adsorbent. In addition, ceramides having a *trans*-double bond in position 4 of the long-chain base are also separable on silica gel impregnated with sodium borate (saturated compounds migrate ahead of unsaturated) although the reason for the effect is not understood.[364] The long-chain base and fatty acid constituents may each contain up to two double bonds so TLC with silica gel impregnated with silver nitrate (5 per cent, w/w) can also be used to good effect in separations of ceramide species.

From a comprehensive study of a wide range of synthetic ceramides on these various layers, Karlsson and Pascher[257] were able to recommend a programme for the analysis of natural mixtures although they recognise that, as the components of natural ceramides vary greatly in chain length, bands may tend to spread more than is the case with pure model compounds. Four groups of ceramide can be separated on the arsenite layers with chloroform–methanol (95:5, v/v) as the solvent system;

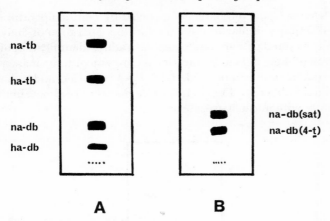

Fig. 8.8. Thin-layer chromatography of ceramides on layers of silica gel G impregnated with 2 per cent sodium arsenite (plate A) and 2 per cent sodium borate (plate B) using a solvent system of chloroform-methanol (95:5, v/v).[257] Abbreviations: na, normal fatty acid; ha, hydroxy fatty acid; db, dihydroxy base; tb, trihydroxy base; db (sat), saturated dihydroxy base; db (4-*t*), $\triangle^4$-*trans*-unsaturated-dihydroxy base.

dihydroxy base-normal acid and dihydroxy base-hydroxy acid as illustrated in Fig. 8.8 (plate A). Derivatives of dihydroxy bases isolated in this manner can then be separated on borate impregnated layers using the same solvent mixture for development, according to whether they contain long-chain bases with *trans*-double bonds in position 4 or not (Fig. 8.8, plate B). Finally all the components isolated thus far can be further separated into groups according to the number of double bonds (almost entirely of the *cis*-configuration) of the component fatty acids together with the number of *cis*-double bonds (*trans*-double bonds have comparatively little effect) in the long-chain bases on silver nitrate impregnated layers. Better results are obtained in this instance if the free hydroxyl groups are first acetylated with acetic anhydride and pyridine (see Chapter 4) when separations of the kind illustrated in Fig. 8.9 are obtained; chloroform–benzene–acetone (8:20:10, by vol.) is then a suitable solvent system for the development.

Ceramides separated in this way can ultimately be resolved in the form of the acetate or trimethylsilyl ether derivatives according to the chain lengths of the aliphatic components by high-temperature GLC under conditions similar to those used in the separation of diglyceride acetates. As standards are not available readily, identification of peaks may present some difficulty unless a gas chromatography–mass spectrometer combination can be used.[80, 183, 185, 456, 457] Ceramides that have not been simplified by TLC in the above manner are almost impossible to identify without such assistance.

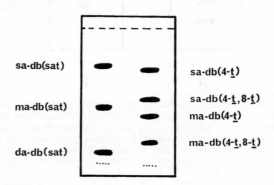

FIG. 8.9. TLC of ceramide acetates on layers of silica gel G impregnated with 5 per cent silver nitrate with a developing solvent of chloroform–benzene–acetone (80:20:10, by vol.).[257] Abbreviations: sa, saturated fatty acids; ma, monoenoic fatty acids; da, dienoic fatty acids; db (sat), saturated dihydroxy-bases; db (4-*t*), △⁴-*trans*-unsaturated dihydroxy-bases; db (4-*t*, 8-*t*) △⁴-*trans*, △⁸-*trans*-diunsaturated dihydroxy-bases.

Intact monoglycosyl ceramides (cerebrosides) have been subjected to high-temperature GLC after conversion of the free hydroxyl groups of the hexose and aliphatic residues to the trimethylsilyl ether derivatives. Compounds are separated according to the combined chain lengths and number of hydroxyl groups in the long-chain base and fatty acid components but the GLC recorder traces obtained are complex and components are not readily identified unless the column effluent is passed into a mass spectrometer.[184] Column conditions similar to those used for triglyceride analysis are employed.

CHAPTER 9

Enzymatic Hydrolysis of Lipids

A. Introduction

Ester bonds to all three positions of the glycerol molecule are so similar in chemical reactivity that there is no simple chemical procedure for hydrolysing one or other of them selectively. Most living organisms, however, have developed lipolytic enzyme systems that are able to distinguish between various positions of glycerol or certain types of bonds in specific lipids and in many cases these enzymes can be isolated and used in simple *in vitro* incubations as an aid in structural analyses of lipids. A lipase (EC. 3.1.1.3), prepared from pig pancreas extracts, for example, is used to hydrolyse quite specifically the fatty acids esterified to the primary positions of triglycerides (see Fig. 9.1) leaving a 2-monoglyceride, the fatty acid composition of which accurately reflects that of position 2 in the original triglyceride. While the same enzyme can be used in an analogous manner with more complex glycerolipids, it is usually technically simpler to hydrolyse glycerophosphatides with the enzyme, phospholipase A_2 (EC. 3.1.1.4), available in a highly active form from snake venoms, which specifically hydrolyses the ester bond in position 2 of natural glycerophosphatides (Fig. 9.1). Such enzymes are useful not only for determining the manner in which fatty acids are distributed among the positions of intact lipids but also for calculating the proportions of various positional isomers of molecular species of lipids separated by the procedures described in Chapter 8.

Other enzymes exist which cleave the glycerol–phosphorus or phosphorus–nitrogenous base bonds in glycerophosphatides, i.e. phospholipase C (EC. 3.1.4.3) and phospholipase D (EC. 3.1.4.4) respectively (Fig. 9.1) and are useful in preparing derivatives of complex lipids

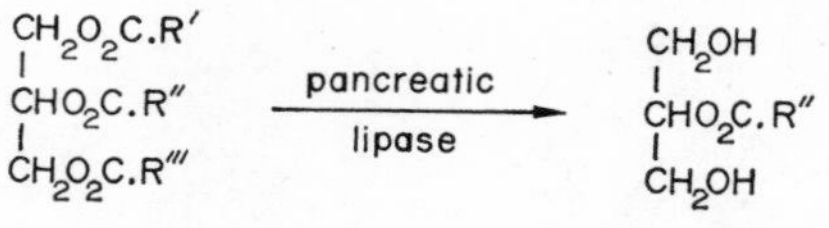

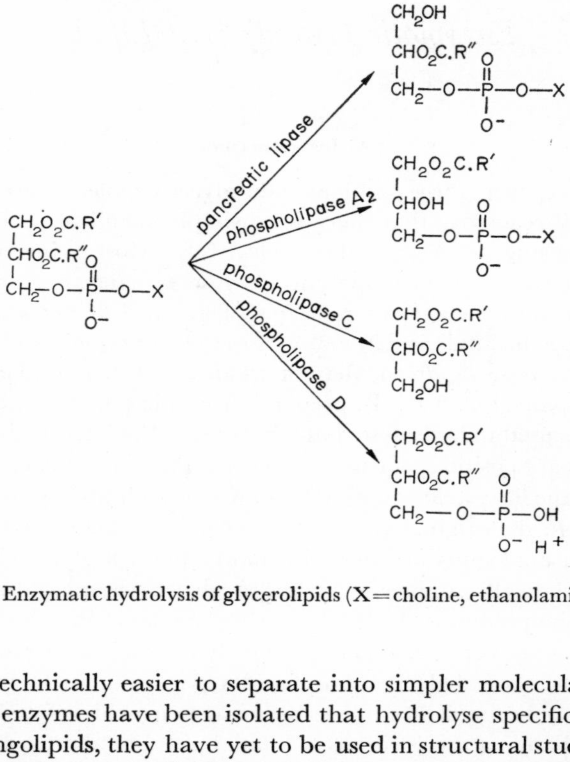

FIG. 9.1. Enzymatic hydrolysis of glycerolipids (X = choline, ethanolamine, etc.).

that are technically easier to separate into simpler molecular species. Although enzymes have been isolated that hydrolyse specific bonds in glycosphingolipids, they have yet to be used in structural studies. Some of these enzymes are available commercially; others, particularly those of microorganisms, can be prepared in a sufficiently pure state for lipid structural studies by simple chemical procedures provided that some microbiological expertise or advice is available to the analyst.

In enzymatic structural studies of lipids, precautions must always be taken to minimise the effects of autoxidation (see Chapter 3).

B. Positional Distribution of Fatty Acids in Triglycerides

1. *Pancreatic lipase hydrolysis*

A lipase (EC. 3.1.1.3) that is completely specific for the primary ester bonds of glycerides is present in crude pig (and other animal) pancreas extracts. Although many other enzymes, including other lipolytic enzymes, and significant amounts of complex lipids are present, such preparations are available commercially at low cost and the impurities do not interfere with structural studies of triglycerides. The properties of the enzyme have recently been reviewed by Jensen[243] and Litchfield.[306] All straight-chain saturated fatty acids of whatever chain length are apparently hydrolysed from the primary positions at approximately the same rate, but the ester bonds of long-chain polyunsaturated fatty acids such as 22:6 (*n*-3), *trans*-3-hexadecenoic acid and phytanic acid to glycerol are hydrolysed more slowly, possibly as a result of steric hindrance caused by the proximity of functional groups to the ester bonds. Unfortunately, the enzyme hydrolyses triglyceride molecules that contain short-chain fatty acids more rapidly than molecules that contain only longer-chain fatty acids, e.g. with a triglyceride such as 1-butyro-2,3-dipalmitin, both the fatty acids on the primary positions are hydrolysed at the same rate but faster than from a triglyceride such as tripalmitin.[243] However, when the enzyme is used on triglycerides that contain a more normal range of fatty acid components, little fatty acid specificity is evident.

Calcium ions are essential and bile salts are helpful for the reaction and it is necessary that the triglycerides be well dispersed as they must be in a micellar form for hydrolysis to occur. For this reason, methyl oleate[38] or hexane[63] are sometimes added as carriers to solid highly saturated fats. In triglyceride structural studies, the concentrations of the various cations, bile salts and the enzyme, the pH of the reaction medium and the temperature are adjusted so that the optimum degree of hydrolysis is obtained (50–60 per cent) in the shortest possible time. Only the 2-monoglyceride formed need be isolated and transesterified so that the fatty acid composition of position 2 of the triglyceride can be determined by gas chromatography (any 1-monoglycerides found are products of acyl migration from 2-monoglycerides). Although diglycerides produced as intermediates in the enzymatic hydrolysis may not be those expected of a random reaction, the fatty acid composition of the monoglycerides rarely appears to deviate far from the true result in practice.

A semimicro method developed by Luddy *et al.*[316] has been widely used and was found to give the best results of a number of published procedures.[21] As the activities of enzyme preparations from different sources may vary, it is advisable to test the method with some natural triglycerides, e.g. lard or corn oil, as substrate to ensure that hydrolysis proceeds at a satisfactory rate.

"Tris(hydroxymethyl)methylamine (tris) buffer (1 M; pH 8·0; 1 ml), calcium chloride solution (2·2 per cent, 0·1 ml) and a solution of bile salts (0·05 per cent; 0·25 ml) are added to the triglycerides (up to 5 mg) in a stoppered test-tube and the whole is allowed to equilibrate at 40°C in a water-bath for 1 min before the pancreatic lipase preparation (1 mg) is added. The mixture is shaken vigorously at this temperature by means of a mechanical shaker for 2–4 min until the desired degree of hydrolysis is attained when the reaction is stopped by the addition of ethanol (1 ml) followed by 6 N hydrochloric acid (1 ml). The solution is extracted three times with diethyl ether (10 ml portions), with centrifugation if necessary to break any emulsions, and the solvent layer washed twice with distilled water (5 ml portions) and dried over anhydrous sodium sulphate. On removal of the solvent, the products are separated by TLC on silica gel G layers (0·5 mm thick on 20 × 10 cm glass plates) with hexane–diethyl ether–formic acid (80:20:2 by vol.) as solvent system. After spraying with 2′,7′-dichlorofluoroscein solution, the monoglyceride band is identified (it is normally the bottom band just above the origin) and recovered from the adsorbent by elution with chloroform–methanol (95:5, v/v) for transesterification or is transesterified directly on the adsorbent. No more than the minimum delay between stages is advisable."

The fatty acid composition of the monoglyceride fraction, which accurately represents that of position 2 of the original triglyceride, is ultimately determined by gas chromatography and it is customary to present results, as in most lipid structural studies, in moles per cent. On the other hand, the free fatty acids released may not be similar to the mean composition of fatty acids originally present in positions 1 and 3 of the triglycerides as

hydrolysis may not be completely random and there may be contamination by fatty acids liberated from lipids endogenous to the enzyme
preparation or by fatty acids released from position 2 after they have
migrated to the primary positions. The mean composition of each fatty
acid in positions 1 and 3 can, however, be calculated from the concentration of each component in the intact triglyceride and in position 2 by
means of the relationship

$$[\text{positions 1 and 3}] = \frac{3 \times [\text{triglyceride}] - [\text{position 2}]}{2}$$

although it is, of course, subject to the cumulative errors of the analyses.

Pancreatic lipase hydrolysis can also be used to determine the fatty
acid composition of each position in diglycerides and diglyceride acetates, but it is necessary to saturate the incubation medium with diethyl
ether for accurate and reproducible results.[292] 1-Alkyl-2,3-diacyl-
glycerols can also be subjected to hydrolysis with the enzyme and the
product, a 1-alkyl-2-acyl-glycerol, isolated and the fatty acid composition of position 2 determined.[508] With both the above types of compounds, as was the case with triglycerides, the free fatty acids released
do not contain only components that were present in the primary
positions, the compositions of which must again be calculated from the
compositions of the intact lipid and of the 2-acyl hydrolysis products.
Certain other lipases related to pancreatic lipase (see section C2 below)
may be more suitable for these compounds.

Pancreatic lipase hydrolysis can be used to determine the fatty acid
composition of position 2 of most triglycerides containing a normal
range of fatty acids or fatty acids with polar or non-polar substituents
remote from the ester linkage. Estolides can also be hydrolysed by the
enzyme and only the glycerol-fatty acid bond is broken while estolide
linkages remain intact.[88, 357] The method is not suitable for triglycerides
containing long-chain polyunsaturated fatty acids in the primary
positions, such as fish oils,[57] and is of limited value for triglycerides
containing significant quantities of short-chain fatty acids, such as
ruminant milk fats.[244] Fortunately, a chemical method has been developed that may be appropriate to these circumstances.[609]. In this triglycerides are reacted with a Grignard reagent that liberates fatty acids
(converting them to tertiary alcohols) with formation of partial glycer-

ides, provided that the reaction is not allowed to go to completion. The various products are isolated by TLC on layers of silica gel G impregnated with boric acid and, although some acyl migration inevitably occurs, the amount of each fatty acid in the β-position can be calculated with acceptable accuracy from the concentration of each component in the original triglycerides and in the α,β-diglycerides, formed in the reaction, by means of the relationship

$$[\beta\text{-position}] = 4 \times [\alpha,\beta\text{-diglycerides}] - 3 \times [\text{triglycerides}]$$

The result is of course subject to the cumulative error of the analyses. The Grignard reaction is discussed in greater detail in the next section.

2. *Stereospecific analysis of triglycerides*

No lipolytic enzyme has yet been discovered that distinguishes between positions 1 and 3 of a triacyl-*sn*-glycerol. None the less, a number of ingenious stereospecific analysis procedures for determining the composition of fatty acids esterified to positions 1, 2 and 3 of L-glycerol in triglycerides have been devised, principally by Brockerhoff who has recently reviewed the available methods.[66] Such procedures are also discussed at length in Litchfield's book.[306] Most methods require the preparation of partial glycerides that can be chemically phosphorylated and reacted with a stereospecific lipase or are phosphorylated by stereospecific enzymes. In the most valuable approach[62, 63] (Fig. 9.2), α,β-diglycerides (an equimolar mixture of the 1,2- and 2,3-*sn*-isomers) are prepared by reacting the triglyceride with a Grignard reagent and are converted synthetically to phenylphosphatides by reaction with phenyl-dichlorophosphate and pyridine. The phosphatides are then hydrolysed by the phospholipase A of snake venom which hydrolyses only the "natural" 1,2-diacyl-*sn*-glycerophosphatide (see below). The products, isolated by TLC, are a lysophosphatide that contains the fatty acids originally present in position 1, free fatty acids released from position 2 and the unchanged "unnatural" 2,3-diacyl-phosphatide. After transesterification, the fatty acid composition (in moles per cent) of each product is determined by gas chromatography. In addition, the composition of position 2 can be determined independently by pancre-

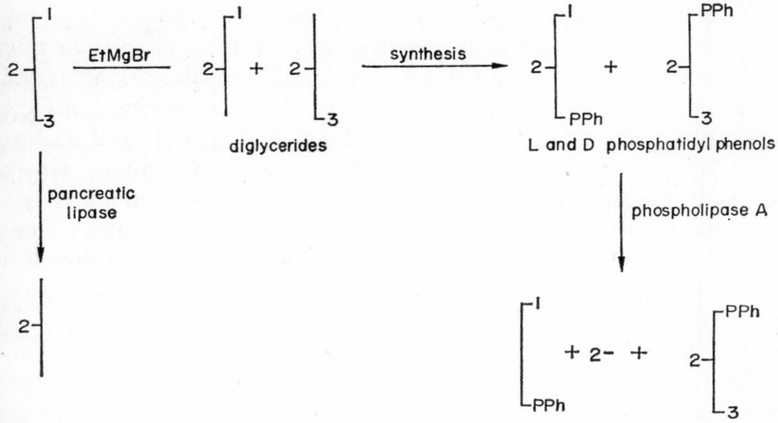

Fig. 9.2. Stereospecific analysis of triglycerides by the procedure of Brockerhoff[62] (PPh = phosphoryl phenol).

atic lipase hydrolysis. Only the fatty acid composition of position 3 is not determined directly but the amount of each fatty acid in this position can be calculated from the analysis of the original triglyceride and those of positions 1 and 2 or from the analysis of the unchanged 2,3-diacyl-phosphatide and that of position 2, i.e. for each component

[position 3] = 3 × [triglyceride] − [position 1] − [position 2]
[position 3] = 2 × [2,3-diacyl-phosphatide] − [position 2]

The key to success with the procedure lies in the preparation of the intermediate diglycerides, which must be generated in a random manner so that the fatty acid compositions of the various positions are the same as in the original triglycerides, i.e. there must be no selectivity for specific fatty acids or fatty acid combinations and a minimum of acyl migration during their formation. Hydrolysis with pancreatic lipase was originally used for the purpose, but is not always suitable as it exhibits specificity for molecules containing certain fatty acid combinations. Therefore, chemical hydrolysis is now preferred and Grignard reagents are used for the purpose as they have no fatty acid specificities and cause less acyl migration than other chemical methods.[609] Methyl magnesium bromide was originally used, but has largely been replaced by ethyl magnesium bromide which reacts rapidly and randomly with the ester bonds of triglycerides converting the fatty acids to tertiary alcohols and, if the reaction is stopped when it is between 30 and 70 per cent complete, a complex mixture of partial glycerides including α,β- and α,α'-(or 1,3-*sn*-)-diglycerides. Unfortunately, some acyl migration always occurs during the reaction and the 1,3-diglycerides contain 6 to 10 per cent of fatty acids that have migrated from position 2, but although position 2 of the α,β-diglycerides is contaminated by up to 4 per cent of fatty acids that have migrated from positions 1 and 3, the fatty acid compositions of the primary positions appear to be identical to those of the original triglycerides.[96] The α,β-diglycerides required for the procedure must not come in contact with polar solvents or be heated and are isolated *immediately* by TLC on layers of silica gel G impregnated with boric acid for conversion without delay to phosphatidyl phenols, which can also be purified by TLC although this does not appear to be essential. Although 1,2-diacylphosphatidyl phenols do not react as rapidly with phospholipase A as most other glycerophosphatides (see section C1 below for more detailed discussion), they are hydrolysed completely by prolonged treatment with the enzyme.

The following modification of Brockerhoff's procedure can be used to determine the composition of fatty acids esterified to each of positions 1, 2 and 3 of a triglyceride provided 10–40 mg of material is available.[96, 97]

"Triglycerides (40 mg) are dissolved in dry diethyl ether (2 ml) and freshly prepared ethyl magnesium bromide in dry diethyl ether (1 ml, 0·5 M) is added. The mixture is shaken for 1 min then glacial acetic acid (0·05 ml) followed by water (2 ml) are added to stop the reaction and the products are extracted with diethyl ether (three 10 ml portions). The ether extract is washed first with aqueous potassium bicarbonate (2 per cent, 5 ml), then with water (5 ml) and is dried over anhydrous sodium sulphate. After removal of the solvent at ambient temperature, the required α,β-diglycerides are obtained by preparative TLC on silica gel G layers containing 5 per cent (w/w) boric acid (0·5 mm thick on a 20 × 20 cm plate) developed in hexane–diethyl ether (1:1, v/v). Bands are visualised under UV light after spraying with aqueous Rhodamine 6G (0·01 per cent, w/v) (see plate A, Fig. 9.3) and that containing the α,β-diglycerides is scraped into a small chromatographic column and eluted with diethyl ether (100 ml.). Small amounts of boric acid are also eluted but do not interfere with subsequent steps.

"The diglycerides (yield generally 6–7 mg) in diethyl ether (1 ml) are reacted with phenyldichlorophosphate (0·2 ml) in pyridine (1 ml) at room temperature. After 90 min, more pyridine (2 ml) is added followed by water (1 ml) dropwise. The products are transferred to a separating funnel with chloroform (30 ml), methanol (30 ml) and triethylamine (0·2 ml) when water (25 ml) is added and the whole thoroughly mixed. The lower layer is collected and the solvent removed under reduced pressure. No delays are permissible between steps until this stage is reached.

"The phospholipid is dissolved in diethyl ether (3 ml) and 0·5 M trisbuffer (0·5 ml; pH 7·5; 2×10^{-3} M with respect to calcium chloride) and *Ophiophagus hannah* snake venom (1 mg) are added and the mixture shaken overnight. *Iso*-propanol (10 ml) and acetic acid (0·2 ml) are then added and the whole taken to dryness. The products in a little chloroform–methanol (2:1, v/v) are applied in a band to a 20×20 cm TLC plate coated with silica gel G (0·5 mm thick) which is developed in hexane–diethyl ether–formic acid (50:50:1, by vol.). The top third of the plate is sprayed with 2′,7′-dichlorofluorescein in methanol (0·1 per cent, w/v) and the free fatty acid band identified and scraped off for methylation (see Chapter 4). The plate is then redeveloped with a solvent system of chloroform–methanol–14 M ammonia (90:8:2, by vol.) up to the level of the previous spray. The lysophosphatide and the unchanged 2,3-diacyl-phosphatide bands are detected by spraying with Rhodamine 6G (0·01 per cent, w/v) (see plate B, Fig. 9.3) and are scraped off and transesterified for GLC analysis."

Although the procedure requires two synthetic steps, two enzymatic hydrolyses (including a separate pancreatic lipase hydrolysis), two thin-layer steps and five transesterifications and GLC analyses, consistent and reliable results can be obtained with practice. Analysis of the lysophosphatide in the final stage of the procedure gives trustworthy results for the fatty acid composition of position 1, but that of position 2 can

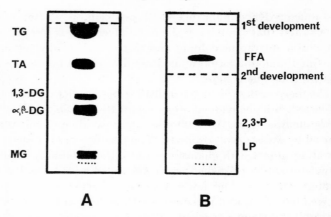

Fig. 9.3. Stereospecific analysis of triglycerides. Plate A, TLC separation of Grignard deacylation products on layers of silica gel G impregnated with 5 per cent (w/w) boric acid; solvent system, hexane–diethyl ether (1:1, v/v). Plate B, TLC separation of products of phospholipase A hydrolysis on layers of silica gel G. Solvent system—1st development, hexane–diethyl ether–formic acid (50:50:1, by vol.). Solvent system—2nd development, chloroform–methanol–14 M ammonia (90:8:2 by vol.). Abbreviations: TG, triglycerides; TA, tertiary alcohols; DG, diglycerides; MG, monoglycerides; FFA, free fatty acids; 2,3-P, 2,3-diacyl-phosphatide; LP, lysophosphatide.

usually be determined with greater accuracy by means of an independent hydrolysis with pancreatic lipase (with the exception of fish oils or milk fats), rather than by analysis of the free fatty acids released by phospholipase A. The result for position 3 can be calculated by the two procedures described above (the first method is generally favoured) and is subject to the cumulative experimental error of the other analyses but is probably accurate to 2–3 per cent (absolute) for major components. Analyses should not be accepted unless the results for major components in positions 2 and 3, determined by both methods, agree within 4 per cent (absolute).

Ethyl magnesium bromide cannot be used to generate diglycerides when the triglycerides contain fatty acids with polar functional groups such as estolide bonds[88] that may also react with the reagent. Pancreatic lipase must then be used for the purpose and the errors potentially associated with this method must be accepted.

The main drawback of this approach to stereospecific analysis of tri-glycerides is that the fatty acid composition of position 3 is not determined directly so that errors in the results for minor fatty acid components in this position may be considerable. Indeed small negative values are sometimes obtained and small positive values may arise in calculations when the component is not present in the position at all. The author (unpublished work) has attempted to overcome this problem by reacting the unchanged 2,3-diacyl-phosphatide, produced in the final stage of the procedure, with pancreatic lipase to release the fatty acids from position 3 for analysis. Unfortunately, the best enzyme preparations available still contained endogenous lipid that could not readily be removed so that the free fatty acid fraction was contaminated. Brocker-hoff devised an alternative approach (Fig. 9.4, scheme A)[65] in the hope of resolving the difficulty. 1,3-Diglycerides, prepared by reaction with the Grignard reagent, are isolated in this procedure and are phosphoryl-ated and reacted with phospholipase A as before. The fatty acids in position 1 are released leaving a lysophosphatide that contains the fatty acids originally present in position 3. Unfortunately, 1,3-diglycerides are contaminated by significant amounts of isomerised α,β-diglycerides, as discussed above, so that positions 1 and 3 may contain 6–10 per cent of fatty acids originally present in position 2 and the method is much less precise than the first described.

One other important analytical method has been described by Lands and his colleagues (Fig. 9.4, scheme B).[297] α,β-Diglycerides are again prepared (pancreatic lipase hydrolysis was used for the purpose in the original procedure, but reaction with ethyl magnesium bromide would probably now be considered more suitable) and are phosphorylated enzymatically by a diglyceride kinase from *E. coli* that is completely stereospecific for 1,2-diacyl-*sn*-glycerols. The fatty acid composition of the phosphatidic acid formed is determined and that of position 2 is obtained separately by pancreatic lipase hydrolysis so that the results for the concentrations of each acid in positions 1 and 3 can be calculated, i.e.

[position 1] $= 2 \times$ [phosphatidic acid] $-$ [position 2]
[position 3] $= 3 \times$ [triglyceride] $- 2 \times$ [phosphatidic acid]

The procedure has been little used, probably because the relevant enzyme is not available commercially, but merits greater notice as the

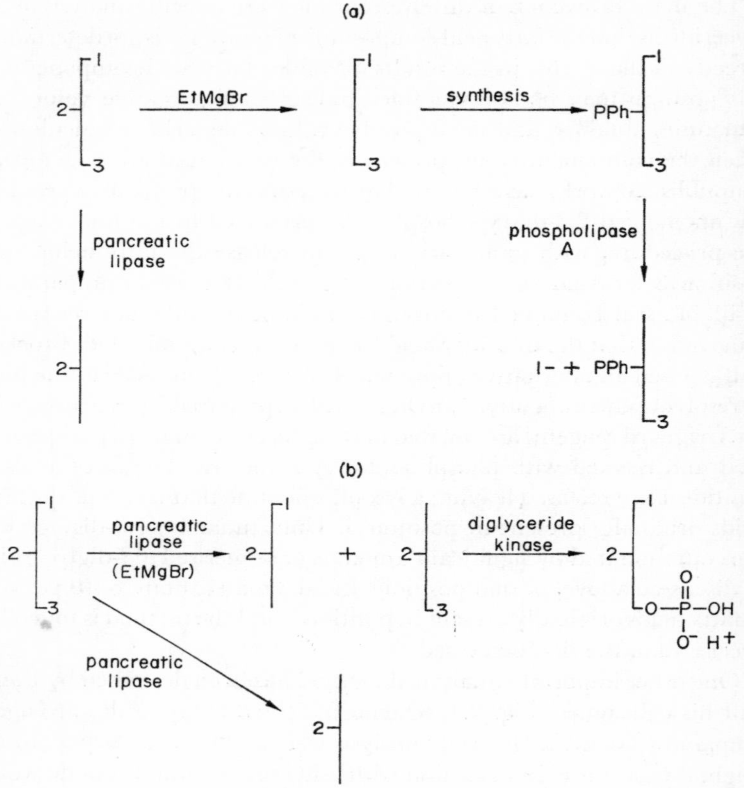

FIG. 9.4. Stereospecific analysis of triglycerides. Scheme A—Brockerhoff's second procedure[65] (PPh=phosphoryl phenol). Scheme B—Lands' procedure.[297]

1,2-diacyl-phosphatide that is isolated is also an important intermediate in triglyceride biosynthesis.

It should be recognised that an analysis of a natural triglyceride which indicates that positions 1 and 3 have identical fatty acid compositions does not necessarily imply that there is an absence of specific stereoisomers and simpler molecular species must be subjected to stereospecific analysis before such conclusions can be justified unequivocally.

C. Enzymatic Hydrolysis of Complex Lipids

1. *Phospholipase A₂*

Phospholipase A_2 (EC. 3.1.1.4) hydrolyses the ester bond in position 2 of glycerophosphatides (see Fig. 9.1) releasing the fatty acids in this position. The products, free fatty acids and a lysophosphatide containing the fatty acids of position 1, are isolated for analysis so that the distribution of fatty acids in both positions of the glycerol moiety is determined. Although the enzyme is found in many animal tissues, the most important sources for use in structural studies are snake venoms. That of *Crotalus adamanteus* or *C. atrox* has been used most often, but is now being supplanted by that of *Ophiophagus hannah*.[383] The mode of action of the enzyme has been reviewed.[547] It is completely stereospecific for L-glycerophosphatides and will react with most natural compounds of this type including phosphatidyl choline, phosphatidyl ethanolamine, phosphatidyl inositol, phosphatidyl glycerol, phosphatidyl serine, phosphatidic acid and cardiolipin,[547] but not with polyphosphoinositides. It also reacts with the phosphatidyl phenols that are intermediates in stereospecific analysis of triglycerides and with synthetic phospholipids containing the phosphorus group in position 2 in which instance the fatty acids in position 1 of L-glycerol are released. Although the phospholipase A in some venoms exhibits specificity for certain fatty acid combinations,[353,383] this need not be troublesome in structural studies provided that the hydrolysis is taken to completion. Fortunately, the enzyme in *O. hannah* does not exhibit such specificities and is especially suited to structural studies.[383] Calcium ions are essential for the reaction which is also stimulated by diethyl ether.

It is necessary to modify the reaction conditions somewhat according to the nature of the phospholipid to be hydrolysed; phosphatidyl cholines have been most often studied in this manner and the following procedure, essentially that of Robertson and Lands,[439] is widely used for the purpose.

"*Ophiophagus hannah* snake venom (2·5 mg) is dissolved in tris buffer solution (0·5 ml; pH 7·5) containing calcium chloride (4 mM). Phosphatidyl choline (5 to 50 mg) is dissolved in diethyl ether (2 ml) and 100 μl of the snake venom solution is added and the whole shaken vigorously for 1 hr. The mixture is washed into a conical flask with methanol (10 ml) followed by chloroform (20 ml) and the solution dried over anhydrous sodium sulphate. After filtering and removing the solvent *in vacuo*, the products are applied to a silica gel G TLC plate (20 × 10 cm; 0·5 mm thick layer) which is developed first in hexane–diethyl ether–formic acid (60:40:2 by vol.). The top half of the plate is sprayed with 2',7'- dichlorofluorescein solution and the free fatty acid band scraped off and esterified. When the plate is dry, it is redeveloped in chloroform–methanol–acetic acid–water (25:15:4:2, by vol.) and the lysophosphatidyl choline is detected and transesterified for GLC analysis (the plate should be similar to plate B, Fig. 9.3, except that there will be no unchanged diacyl-phosphatide if the reaction has gone to completion)."

The accuracy of the procedure should be checked by summing the results for the concentrations (moles per cent) of each fatty acid in positions 1 and 2, dividing by two, and comparing this quantity with the analysis for each component in the original phosphatidyl choline; the two results should agree within 1 per cent (absolute) for each component.

Difficulties are frequently reported when the procedure is used with other glycerophosphatides, particularly with phosphatidyl ethanolamine, but the use of borate buffers appears to alleviate the problem.[388, 550] The author has found the following procedure satisfactory for phosphatidyl ethanolamine and phosphatidyl inositol.[550]

"Phosphatidyl ethanolamine (8 mg) is dissolved in diethyl ether (2 ml) and 0·1 M borate buffer (0·25 ml; pH 7), 0·01 M calcium (0·125 ml), water (0·125 ml) and *O. hannah* snake venom (1 mg) are added. The mixture is shaken thoroughly for 1 hr then worked up as above."

It may be necessary to prolong the reaction somewhat with more acidic lipids and to vary the polarity of the second solvent system in the TLC separation according to the polarity of the lipid hydrolysed. With phosphatidyl serine, it is necessary to render the buffer solution alkaline with 0·1 N potassium hydroxide solution to enable the reaction to go to completion.[605] The accuracy of the procedure should always be checked in the manner described above for phosphatidyl choline.

Phospholipase A_2 from *Crotalus* species hydrolyses diacyl phosphatides much more rapidly than the analogous ether compounds[567] and has been used to obtain fractions enriched in the latter (see Chapter 7). When ether-containing phospholipids are present in a sample, it is therefore necessary to prolong the hydrolysis time to ensure complete reaction if the fatty acid composition of position 2 of the mixture is required. The enzyme in *Naja naja* venom, however, hydrolyses alkyl phosphatides at the same rate as the diacyl compounds but twice as rapidly as it hydrolyses alkenyl phosphatides.[567] Although the phospholipase A of snake venom does not hydrolyse the secondary ester bond in di- and triphosphoinositides, a related enzyme is found in bee venom that performs this function and it has been used for the purpose in structural studies.[538]

The existence of a phospholipase A_1 that hydrolyses the primary ester bond of glycerophosphatides has been demonstrated in rat liver and pancreas but may in fact be similar to or the same enzyme as pancreatic lipase.

2. *Pancreatic lipase and related enzymes*

(i) *Glycerophosphatides*

Pancreatic lipase will hydrolyse the ester bond to position 1 of glycerophosphatides (Fig. 9.1) leaving a lysophosphatide containing the fatty acids of position 2. The following method is generally satisfactory.[497, 498]

"The phospholipids (10 mg), bile salts (3 mg) and bovine serum albumin (4·5 mg) are dispersed by ultrasonification in 0·1 M borate buffer (pH 8·0; 1 ml) made 5×10^{-3} M with calcium chloride when pancreatic lipase (10 mg) is added. The mixture is shaken vigorously for 0·5 to 6 hr (determined by experiment) until hydrolysis is complete. The mixture is extracted three times with chloroform–methanol (7:3, v/v) and the lysophosphatide is obtained for analysis, after removing the solvent, by TLC on silica gel G layers as described above for phospholipase A hydrolyses."

Unfortunately, unless highly purified enzyme preparations are used (and they are not available commercially), the free fatty acids released will

K

be contaminated by fatty acids endogenous to the enzyme preparation and possibly by some that have migrated from position 2 to position 1 before being hydrolysed. Only the result for position 2 is then meaningful, but that for position 1 can of course be obtained by difference as the composition of the intact lipid is known. The preparation of the pure pancreatic enzyme is a tedious process with a large number of steps.[474, 556] Pancreatic lipase hydrolysis of phospholipids has been suggested as a means of obtaining concentrates of alkyl- and alkenyl- components (see Chapter 7) and it is also used in the semi-synthetic preparation of 2-acyllysophosphatides.[498]

A lipase with the same positional specificity as pancreatic lipase has been isolated from the mould *Rhizopus arrhizus* and may be more suitable for the hydrolysis of phospholipids.[497, 498] It is used in a similar manner to pancreatic lipase except that it has a pH optimum of 6·5. Regrettably, it is not yet available commercially.

Neither enzyme has been much used in structural studies of phospholipids to date as they do not offer significant advantages over phospholipase A_2 with the more conventional glycerophosphatides. On the other hand, they may be useful with semi-synthetic non-polar phosphatides such as dimethylphosphatidic acid, that is frequently prepared from the more common phospholipids for separation into simpler molecular species (see Chapter 8).

(ii) *Glycosyl diglycerides*

Pancreatic lipase also hydrolyses the ester bond to position 1 of glycosyl diglycerides with production of free fatty acids and a glycosyl monoglyceride containing the fatty acids of position 2.[131] Both products can be isolated by TLC and esterified for GLC analysis. For the reasons given above, only the fatty acid composition of position 2 is in general determined directly while that of position 1 is calculated by difference from the known composition of the intact lipid. The following procedure is based on that of Noda and Fujiwara.[381]

"Pancreatic lipase (10 mg), 45 per cent calcium chloride solution (0·03 ml), 1 per cent sodium deoxycholate solution (0·01 ml) and the glycosyl diglyceride (15 mg) are dispersed by vigorous shaking in

0·2 M tris buffer (pH 7·6; 1 ml). The mixture is incubated at 40°C for 1 hr with constant shaking before it is acidified to pH 3 with 1 M sulphuric acid. The products are extracted with chloroform–methanol as in the phospholipase A procedure above and are separated by TLC on silica gel G layers in a similar two step system. Free acids are obtained in the top half of the plate with hexane–diethyl ether–formic acid (60:40:2, by vol.) while the monoacyl products are then separated by means of a further development with a solvent system of chloroform–methanol–acetone (80:16:4, by vol.; monogalactosyl monoglycerides) or chloroform–methanol–acetic acid–water (80:40:2:3, by vol.; digalactosyl monoglycerides)."

Native mono- and digalactosyl diglycerides have been subjected to hydrolysis in this manner in a number of laboratories,[35, 381, 454] but Arunga and Morrison[30] obtained better results when the carbohydrate moieties were acetylated prior to the reaction.

3. *Phospholipase C*

1,2-Diglycerides are released from glycerophosphatides on digestion with phospholipase C (EC. 3.1.4.3) prepared from the microorganisms *Clostridium welchii* and *Bacillus cereus* (see Fig. 9.1). Enzymes from both sources are commercially available and highly active preparations of that from *B. cereus* can be obtained by ammonium sulphate precipitation from the supernatant fluid used as a growth medium for the microorganisms.[390] If the enzyme is to be used in structural studies, little further purification is necessary. Although there is some evidence that molecular species containing shorter-chain fatty acids are hydrolysed more rapidly than those with longer-chain components, the effect need not be troublesome as the reaction can usually be taken to completion with care. The enzyme of *B. cereus* does not possess an absolute specificity for a phosphate bond in position 3 of L-glycerol as it has been found that it reacts with synthetic phosphatides with the phosphate group in positions 1, 2 and 3.[115]

Phospholipase C from *C. welchii* is used most often for the preparation of diglycerides from phosphatidyl choline and, in exactly the same manner, for the preparation of ceramides from sphingomyelin.[425]

"The phospholipase C of *C. welchii* (1 mg) in 0·5 M tris buffer (pH 7·5; 2 ml), 2×10^{-3} M in calcium chloride, is added to phosphatidyl choline (10 mg) in diethyl ether (2 ml). The mixture is shaken at room temperature for 3 hr when it is extracted (3 times) with diethyl ether. The ether layer is dried over anhydrous sodium sulphate before the solvent is removed at ambient temperature. Pure, 1,2-diacyl-*sn*-glycerols are obtained by preparative TLC on layers of silica gel G impregnated with boric acid (10 per cent, w/w) with hexane–diethyl ether (50:50, v/v) as solvent system. The appropriate band is located with the 2′,7′-dichlorofluorescein spray and is eluted from the adsorbent with diethyl ether (boric acid, which is also eluted, does not interfere with subsequent stages)."

No unnecessary delays are allowable at any stage and the compounds should not be heated or permitted to come in contact with polar solvents otherwise acyl migration may occur and, if molecular species of the diglycerides themselves are required, they should be fractionated immediately. Alternatively, the diglycerides may be acetylated at once with acetic anhydride and pyridine (see Chapter 4 for practical details) as the diglyceride acetates can be stored indefinitely in an inert atmosphere at low temperatures without coming to harm, although they should be purified by preparative TLC before being analysed further. A small portion of the diglycerides or diglyceride acetates should be transesterified in order that its fatty acid composition may be compared with that of the original lipid to ensure that random hydrolysis of molecular species has occurred.

Ceramides, prepared by this method, are comparatively stable but should be purified on thin layers of silica gel G with chloroform-methanol (90:10, v/v) as solvent system.

Although *C. welchii* preparations can also be used to hydrolyse phosphatidyl ethanolamine provided that some lysophosphatidyl choline[533] or sphingomyelin[425] are present to activate the enzyme, better results are obtained with the phospholipase C of *B. cereus*, which can also be used to prepare diglycerides from phosphatidyl serine, phosphatidyl inositol and cardiolipin. The author has found the following method[591, 597] satisfactory for the purpose.

"Phosphatidyl ethanolamine (5 mg) and sphingomyelin (3 mg) are mixed with 0·2 M phosphate buffer (pH 7·0; 0·5 ml) containing 0·001 M 2-mercaptoethanol and 0·0004 M zinc chloride and phospholipase C of *B. cereus* (1 mg) in the above buffer (0·5 ml) is added. The mixture is shaken vigorously for 2 hr at 37°C when the diglycerides produced are extracted and purified as in the procedure immediately above."

The phospholipase C of *Clostridium perfringens* is specific for the diacyl forms of phospholipids yet that of *B. cereus* hydrolyses the 1-O-alkyl forms of phospholipids three times as quickly as the diacyl or alkenyl forms.[567]

A related enzyme, found in brain, has been used to prepare diglycerides from mono-, di- and triphosphoinositides.[220, 540] While hydrolysis does not go to completion, representative diglycerides appear to be obtained although some acyl migration occurs with formation of 1,3-diglycerides, probably because of the low pH optimum of the enzyme. Phosphatidic acid has been dephosphorylated by a phosphatase found in liver mitochondria and microsomes,[202] but the preparation of the enzyme is fraught with difficulty. The acidic phosphatase (EC. 3.1.3.2) from wheat germ will perform the same feat[51] and, as it is readily available commercially, it will probably be favoured in future.

4. *Phospholipase D*

Phosphatidic acid is formed from phosphatidyl choline, phosphatidyl ethanolamine, phosphatidyl serine and phosphatidyl glycerol by reaction with phospholipase D (EC. 3.1.4.4) (see Fig. 9.1), but phosphatidyl inositol is attacked very slowly and cardiolipin not at all by the enzyme. Although the enzyme from vegetable sources is available commercially, highly active preparations can be made with very little effort simply by homogenising cabbage (or brussel sprouts) with an equal volume of water, filtering and centrifuging at 13,000 *g* for 30 min when the supernatant fluid is used directly.[601] Hydrolyses with the enzyme are carried out in the following manner.[601]

"0·1 M sodium acetate (1.25 ml; pH 5·6), 1 M calcium chloride (0·25 ml) and fresh enzyme solution (2 ml) are added to the glycerophosphatide (10 mg) in diethyl ether (1 ml). The mixture is shaken vigorously for 3 hr then brought to pH 2·5 by addition of glacial acetic acid before it is extracted three times with chloroform–methanol (2:1, v/v; 4 ml portions). The lower layer, containing the phosphatidic acid, is taken to dryness and can be purified if necessary by TLC on layers of silica gel G with chloroform–methanol–water (25:10:1, by vol.) as developing solvent (the product is found just below the solvent front)."

Alcohols must be excluded from the reaction medium as the enzyme also catalyses esterification of the phosphate group. Phosphatidic acid formed in the above manner is converted to the dimethyl derivative for separation into simpler molecular species (see Chapters 7 and 8) by reaction with diazomethane in diethyl ether.[601]

"Crude phosphatidic acid in diethyl ether solution (1 ml) is allowed to react with excess diazomethane in diethyl ether for 16–18 hr at room temperature. The solvent is removed in a stream of nitrogen and the dimethylphosphatidate purified by preparative TLC on layers of silica gel G with carbon tetrachloride–chloroform–glacial acetic acid–ethanol (120:80:1:6, by vol.) as developing solvent. The required product has an R_f value of approximately 0·4 to 0·5."

Phospholipase D has also been used to obtain concentrates of alkyl-acyl- and alkenyl-acyl- phospholipids as it reacts much less rapidly with them than with the diacyl forms of a given lipid (see Chapter 7). For example, phospholipase D from carrot and cabbage hydrolyses the 1-O-alkyl and 1-O-alkenyl forms of phosphatidyl choline at one third and one-tenth the rate respectively of the diacyl component.[567]

D. The Use of Enzymatic Hydrolysis in the Determination of Molecular Species of Lipids

None of the methods described in Chapter 8 can be used to separate molecular species of glycerolipids differing in the positional distribution of specific fatty acids; for example, 1-palmityl-2-oleyl- and 1-oleyl-2-palmityl phosphatidyl choline or their derivatives cannot be resolved physically by any method yet developed. If the molecular species containing the mixture of these components can be isolated by chromatographic techniques, however, the proportions of each can be calculated from the results of phospholipase A (phospholipid) or pancreatic lipase (diglyceride acetate) hydrolysis simply by obtaining the ratio of palmitic acid say in each position. Similar considerations apply to the analyses of most other lipids.

One major difficulty arises with molecular species of triglycerides that contain three different types of fatty acid[64] (the so-called "three acid problem") such as saturated, monoenoic and dienoic acids for example. If these are denoted by S, M and D respectively, six different stereo-isomers exist—SMD, SDM, MSD, MDS, DSM and DMS, but stereo-specific analysis can only be used to ascertain the combined amounts of the pairs SMD and SDM, MSD and MDS and DSM and DMS but not the proportions of each isomer in the pair. The problem can be resolved when the stereospecific analysis procedure of Lands *et al.*[298, 494] is used if the 1,2-diacylphosphatides formed in the final stage of the procedure (see section B2 above) are separated by silver nitrate TLC and hydrolysed with phospholipase A. Hammond indeed has shown theore-tically[186] how all the species present in a complex natural triglyceride mixture such as coconut oil can be calculated by applying various stereospecific analysis procedures to the fractions obtained by combina-tions of silver nitrate TLC and high-temperature GLC. The technical difficulties are enormous, however.

CHAPTER 10

The Analysis and Radioassay of Isotopically-labelled Lipids

A. Introduction

The sensitivity of methods for detecting radiotracers is much greater than that of most other chemical and physical procedures and the use of lipids labelled with β-emitters such as ^{14}C, ^3H or ^{32}P has revolutionised the study of dynamic biological processes. An extra dimension is then added to the problem of the analysis of lipids labelled with these isotopes, as both the mass and radioactivity (and thence the specific activity) of individual components of a mixture may have to be determined. In addition, the analyst must be aware of and take precautions against the hazards associated with radiation; this information is available in a number of textbooks. With low levels of ^{14}C- or ^3H-labelled lipids such as are likely to be encountered in most biological experiments, the physical dangers to the operator are negligible but it is none the less a wise precaution to wear protective clothing especially rubber gloves whenever these compounds are handled and it is inadvisable to consume food, smoke or apply cosmetics in the laboratory. Also, contaminated glassware should be segregated from that used for other purposes, all accidental spillages should be noted and scrupulous cleanliness should be observed so that radioactivity is not carelessly transmitted to other parts of the laboratory or carried over to future experiments. Compounds labelled with higher energy β-emitters such as ^{32}P are potentially more hazardous to the operator and should be handled with particular care. Thin layer chromatography of isotopically-labelled lipids should always be performed in fume cupboards to prevent radioactive dust spreading throughout the laboratory. Similarly, the outlets from gas chromatographs that are used in the analysis of radioactive tracers should be led into a fume cupboard or to the outside of the building.

It is not possible to present here a detailed account of the physics of the processes for detection of radioactivity or to discuss the merits of the various systems and it is assumed that the reader has or can readily acquire such knowledge from one of the many textbooks currently available on the subject. Nor is it possible to discuss the availability, commercially or by synthetic and biosynthetic routes, of labelled lipids although this subject has been reviewed elsewhere.[336, 507] It should be noted, however, that both the chemical and radio-purity of all labelled lipids should be checked by appropriate methods (see below) before they are used in biological experiments. Fatty acids labelled with ^{14}C have been stored for over a decade with no radiation-induced decomposition occurring[338] although fatty acids labelled with ^{3}H decompose at a significant rate. It is probably a wise precaution to store labelled lipids at low temperatures in inert solvents so that the energy of radioactive decay is dissipated quickly.

The methods used to separate and estimate isotopically-labelled lipids do not differ markedly from those described above for unlabelled materials and the amounts of the various radioactive compounds recovered from chromatographic processes can be determined by the methods already described provided that the compounds are not altered in such a way that the radioactivity cannot also be measured by appropriate means. For example, radioassay should be performed on a separate aliquot of a sample if the remainder is to be estimated by a destructive procedure such as photodensitometry after charring. In addition, radiodetection systems are extremely sensitive and metabolites may be formed in biological experiments that have, to all intents and purposes, negligible mass but are significantly radioactive so might go undetected if only those compounds that can be seen by visual techniques are separated for counting. It is therefore important to ensure that the yield of radioactivity at the end of a separatory process is close to that applied initially and to check by alternative chromatographic procedures that radioactivity apparently associated with a given compound is due to that compound and not to some other substance that co-chromatographs with it.

With most liquid–solid or liquid–liquid column chromatographic procedures, monitoring of radioactivity is not a problem as, after removal of solvents, the lipids can be dissolved in toluene containing suitable fluors and counted in a liquid scintillation counter. Measurement of radioactivity on thin layer adsorbents or in gas chromatography effluents does present some special difficulties, however, and these are discussed below. Although radioactivity detection systems have been designed specifically for TLC and GLC, liquid scintillation counting can generally be adapted to the purpose and as a result of its great versatility and high sensitivity, this is undoubtedly the most important single counting method (the counting efficiency should always be determined so that counts per min can be converted to disintegrations per min).

In all the following procedures, it is assumed that the analyst will observe all the relevant safety precautions. Precautions should also be taken to minimise the effects of autoxidation (see Chapter 3).

B. Thin-layer Chromatography

Radioactive lipids on thin layer plates can be detected for counting by spraying the edge of the plate with 2′,7′-dichlorofluorescein solution or with water (less sensitive) or by exposing the plate to iodine vapour, marking the bands and allowing the iodine to evaporate off. Alternatively, "cold" standards can be run alongside the labelled compounds and only the former detected by the above methods permitting the position of the latter to be fixed. Bands may then be scraped into small chromatographic columns and simple lipid components eluted from the adsorbent with chloroform–methanol (95:5, v/v) and complex lipids with chloroform–methanol–water (5:5:1, by vol.). The lipids can then be transferred to scintillation vials and, after removal of solvents, dissolved in toluene containing an appropriate mixture of fluors (e.g. 4 g 2,5-diphenyloxazole (PPO) and 0·2 g 1,4-di-(2-(5-phenyloxazolyl))-benzene (POPOP) per litre of toluene) for liquid scintillation counting. It is especially important that all traces of chloroform be removed as this is a powerful quenching agent. With lipids of very high specific activity, it may be necessary to add a "cold" carrier to the eluting solvent to ensure quantitative recovery of radioactivity. Unfortunately, the dyes used to detect lipids on thin layer adsorbents may produce strong chemiluminescent effects enhancing the apparent amount of radioactivity detected, but this phenomenon can be minimised by using the least possible amount of dye and storing the samples in the dark for several hours before counting. If this is done carefully, the whole plate can be sprayed so that any dubiety about the positions of bands is removed. It is generally necessary to determine the mass and radio-activity of components of a mixture in separate analyses.

Non-polar compounds such as methyl esters, cholesteryl esters or triglycerides can be recovered from the scintillant solution for further analyses if necessary by evaporating the solvent and subjecting the fluor–lipid mixture to preparative TLC on silica gel G layers with hexane–diethyl ether (80:20, v/v) as developing solvent; the fluors remain in the bottom half of the plate. Alternatively, the labelled fatty acid moieties can be recovered from scintillant solutions by hydrolysing the mixture with alkali and removing the fluors with the non-saponifiable components. The procedure described in Chapter 4 is suitable or a method devised specifically for this purpose by Howard and Kittinger[231] can be used.

Procedures that involve eluting lipids from TLC adsorbents are tedious and time-consuming and subject to a number of errors so some effort has been devoted to devising means of measuring the radioactivity of compounds while they are still on the plate. Unfortunately, many of these techniques require expensive sophisticated equipment which may not always be available to the analyst. For example, the developed thin-layer plate can be laid on an X-ray photographic film so that the radia-tion produces an image that can be developed and is proportional in intensity to the amount of radioactivity present. The relative intensities of the various bands may be determined by photodensitometry.[50] The TLC plate is unaltered and can then be sprayed with reagents to detect or identify the various components or to determine the masses of each. In an alternative procedure, the developed TLC plate is passed through a radiochromatogram scanner in which radioactivity is measured with two windowless geiger detectors on each side of the plate. The signals from these are transmitted to a recorder so that a continuous trace is obtained illustrating peaks of radioactivity on the plate. Specific spray reagents can then be used to detect and identify components so that the radioactivity peaks can be assigned to specific compounds. Again, the masses of the various components can be separately determined by any

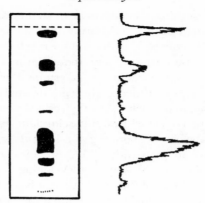

Fɪɢ. 10.1. TLC plate of labelled lipids and radiochromatogram scan.

of the methods described in earlier chapters in order that their specific activities can be calculated. Figure 10.1 illustrates a TLC separation with the radioactivity trace alongside the developed plate. A number of manufacturers supply equipment that can be used for this purpose or for use with paper chromatography and when it is available to the analyst, chromatography on silicic acid impregnated paper may offer advantages over conventional thin-layer chromatography.

Another approach to the problem of radioassay of compounds separated by thin-layer chromatography consists in suspending them on the TLC adsorbent in toluene–fluor and determining the radioactivity of the suspension in a liquid scintillation counter. To ensure that the adsorbent remains in suspension, a gelling agent such as "Cab-o-Sil" (Packard Instrument Co.) may be incorporated into the toluene fluor. Cab-o-Sil is finely divided silica with a very large surface area that forms a thixotropic gel with the sample and prevents self-absorption of activity which occurs if the particles are allowed to settle on the bottom of the counting vial. Webb and Mettrick[571] have critically evaluated procedures of this type and recommended the following method for ¹⁴C-labelled lipids.

"The dried adsorbent, on which the lipids are bound, is placed in a scintillation vial and methanol (1·5 ml) is added. The vial is shaken and left to stand for a few minutes when Cab-o-Sil in toluene–fluor

(4 per cent, w/v; 15 ml) is added. The vial is capped and shaken thoroughly before counting."

The procedure is much more rapid than methods that require prior elution of lipids from the adsorbent, but care must be taken in handling the dry Cab-o-Sil powder which should not be inhaled or allowed to come in contact with the skin. It is necessary to ensure that all the TLC developing solvents, particularly chloroform, are removed prior to placing the sample in the vial. Although the procedure is equally effective with simple and complex ^{14}C-labelled lipids, it is not suited to the radioassay of lipids labelled with ^{3}H as the recoveries may be poor. In this instance, it has been recommended[571] that the silica gel be deactivated with water (1 ml) before being suspended in "Aquasol" (10 ml) for counting. "Aquasol" (New England Nuclear Inc.) and "Unisolve" (Koch-Light Laboratories Ltd.) are proprietary emulsifier-fluors that are much more convenient in use than Cab-o-Sil. Although they are apparently more expensive than the latter, a close study of the economics of various liquid scintillation media may indicate little real difference in the overall cost.

A number of related suspension systems have been described but most appear to be inferior to those described above. Excellent counting efficiencies with weak β-emitters such as ^{3}H have also been obtained by dissolving the silica gel in hydrofluoric acid to prevent lipid adsorption and counting in toluene/triton X-100 scintillation solution.[481] Snyder *et al.*[506, 509] have described an instrument for automatically scraping predetermined areas of the adsorbent from developed TLC plates into scintillation vials for counting in suspension in a manner analogous to that detailed above.

It should again be stressed that all steps in the thin-layer chromatography of radioactive lipids should be performed in an efficient fume cupboard to prevent contamination of laboratories with radioactive dust.

C. Gas–Liquid Chromatography

Isotopically labelled lipids may be subjected to gas chromatography under exactly the same conditions as unlabelled compounds (see Chapter 5 in particular) except that it is doubly important to use silanised supports and catalyst-free liquid phases to minimise retention of components on the columns that would cause the background count to rise and memory effects to become apparent. Facilities for on-column injection are essential and all parts of the instrument that may be exposed to radioactive vapours should be readily replaceable. It is also advisable to cover the whole instrument with a ventilated hood to remove radioactive combustion products.

The combined gas chromatography and radioassay of labelled lipids (especially methyl esters of fatty acids) has been the subject of a number of reviews[238, 258, 476, 484] while one by Karmen[259] in particular treats the subject in much greater depth than is possible here. There are two main approaches to the problem. The first involves adaptation of conventional preparative GLC to collect samples for liquid scintillation counting while, in the second, the GLC effluent is passed through a sensitive radiation detector so that the mass and radioactivity of components are monitored at the same time in a continuous manner.

1. *Radioassay after preparative GLC*

Preparative gas chromatography of methyl esters of fatty acids is described in Chapter 5 and of higher molecular weight compounds in Chapter 8. These procedures, in which a stream splitter diverts part of the column effluent to a flame ionisation detector so that a mass trace is obtained while the bulk of the sample is vented into the collection system, can be used without modification to collect isotopically-labelled components for subsequent assay by conventional liquid scintillation techniques. For example, Watson and Williams[569] have described a conventional system in which components in a GLC effluent are condensed in a trap containing silicone liquid coated on to glass beads (40–60 mesh; 20 per cent, w/w) from which they can be recovered by elution with toluene–fluor for counting. Better than 90 per cent recoveries are claimed. The counting procedure can be simplified, however, if crystals of primary scintillators such as anthracene or *p*-terphenyl coated with 5 per cent (w/v) silicone oil (a slurry of the crystals and the silicone oil in acetone is prepared when the solvent is evaporated) are used to trap components.[259, 260] Traps approximately 4 cm long are filled with this mixture and will collect with great efficiency the material in a given peak issuing from a GLC column. There is then no need to recover the compounds which can be counted simply by placing the trap in a conventional liquid scintillation counter. In this system, ^{14}C is counted with an efficiency comparable to that in solution in a liquid scintillator, but the efficiency of ^3H counting is poor and irreproducible. With compounds labelled with ^3H, therefore, better results are obtained if all the material in the trap (*p*-terphenyl only *not* anthracene) is dissolved in a conventional toluene–fluor for counting. Indeed the whole trap can be placed in the vial with the toluene–fluor as the contents will dissolve while the glass of the trap does not appear to interfere with the scintillation process.

A number of procedures have also been described in which components in the GLC effluent are condensed by bubbling the exit gas into toluene–fluor in a scintillation vial that can later be placed in a liquid scintillation counter for assay. Possibly the most effective of these systems is one devised by Dutton[125] that has been used in a number of laboratories. The outlet of the gas chromatograph is connected via narrow bore stainless-steel tubing, surrounded by a heated aluminium jacket that serves as a heat sink, to a three-way syringe valve. A syringe containing toluene–fluor is attached to one of the remaining outlets while the third outlet is connected again to a length of narrow bore stainless-steel tubing that can be directed into a scintillation vial containing 15 ml toluene–fluor. When samples are to be taken, the GLC effluent is bubbled directly into the scintillation solution and, at the end of the sampling period, the sampling device is flushed with 1 ml of toluene–fluor from the syringe reservoir to minimise tailing of radioactivity before the scintillation vial is changed. The back-pressure associated with the flushing operation causes a response on the mass trace marking the time when the sample was taken. Fractions are collected by reference to particular peaks on the mass trace or better at regular predetermined time intervals for counting. The system can also be automated[536] by arranging to pass toluene–fluor continuously past the heated column exit into a mechanical fraction collector loaded with scintillation vials. In addition, the resolution and sensitivity of the analysis can be considerably improved by combining and averaging the results of several replicate analyses by means of a computer.

Inaccuracies associated with uneven recovery of components can be overcome by an elegant procedure described by Bishop *et al.*[49] Free or combined ³H-labelled fatty acids are converted to their methyl-¹⁴C esters and ¹⁴C-labelled fatty acids to their methyl-³H esters. If the specific radioactivity of the methyl group is known, the ¹⁴C to ³H ratio in the esters separated by GLC gives a direct measure of the specific activity of the fatty acid and quantitative collection is unnecessary. The fatty acids or lipids must be methylated with the minimum amount of labelled methanol for reasons of economy but suitable methods are available.[49]

Techniques of collection followed by counting have several advantages over continuous radioassay systems of which the most significant is that comparatively low levels of radioactivity can be measured by extending the counting time in the liquid scintillation counter. Also, ¹⁴C- and ³H-labelled lipids can be handled with equal facility and labelled fatty acids, in particular, can be recovered from the toluene–fluor if necessary using the methods discussed in the previous section. The main disadvantage of non-automated systems is that the collection of large numbers of fractions is tedious and requires the constant attention of a skilled operator. If the fractions collected are too large, it may not be easy to assign the measured radioactivity to specific peaks in the mass trace and, if the entire column effluent is not collected, some compounds that may unexpectedly contain radioactivity may be missed. Some lipids, such as cholesterol,[259] may not condense with sufficient rapidity in the toluene–fluor and so may be lost.

2. *Continuous monitoring of the column effluent*

The principal advantages of monitoring the effluent of a gas chromatographic column by flow counting techniques are that selective losses are unlikely to occur and that the mass and radioactivity profiles of the components of a mixture can be displayed simultaneously on the same recorder trace. Thus, routine analyses of numbers of samples can be performed. Although many radioactivity detectors can be operated at the temperatures of the gas chromatographic columns, better results are obtained if the effluent is combusted first so that ¹⁴C is counted as ¹⁴CO₂ and ³H as ³H₂ at ambient temperature. If this is not done, radioactive compounds may be adsorbed on the walls of the detector, reducing the amount of radioactivity detected (as such compounds are counted in 2π geometry) and causing "memory" effects with other compounds eluting from the column and generally raising the background radioactivity.

In most commercial radio-gas chromatography systems, the effluent is split as it leaves the column and a portion passes into a flame ionisation detector so that the masses of components are recorded. At the same time, the bulk of the effluent is diluted with cold carbon dioxide and passed into a quartz tube packed with cupric oxide wire maintained at approximately 700°C where organic compounds are converted to carbon dioxide and water. If ^3H is to be assayed also, the effluent is then passed into a column of steel wool, also at 700°C and flushed continuously with hydrogen, where the ^3H_2O is reduced to elemental hydrogen. Finally, the gases are passed through a water trap containing anhydrous magnesium perchlorate into the radioactivity detector.

Although many types of detector have been tried, proportional counters have proved most successful and are used in most commercial instruments. The sensitivity of such detectors is dependent on the time that the substances remain in the detector and comparatively large cells (50–100 ml volume) are often used to obtain the maximum possible radioactivity count. On the other hand if the cell is too large, mixing of adjacent components in the GLC effluent may occur and there will be a loss of resolution (with peak broadening in the radioactivity trace relative to the mass trace). For this reason, some manufacturers use smaller detector cells (approximately 10–20 ml volume) and accept lower counting efficiencies for the sake of the higher resolution that can be attained. With the large cells, a counting efficiency of 80 per cent is possible with ^{14}C, but with smaller cells the efficiency may be as low as 20 per cent depending on the flow rate used. With proper shielding of the detector, the background count can be as low as 2 cpm.[484] Argon is the most effective carrier gas for use with proportional counters and while this is generally diluted with 5 per cent (v/v) carbon dioxide immediately before the combustion train to prevent adsorption effects and to serve as a quencher in the proportional counter, other workers[521] use argon–carbon dioxide (95:5, v/v) as the carrier gas for the entire chromatographic process.

For the visual presentation of results, the detector signal is sent via a ratemeter to one channel of a dual pen recorder, the other channel of which is used to record the mass response from the flame ionisation detector. The radioactivity associated with each component in the sample is then seen directly as in the specimen trace in Fig. 10.2. Some

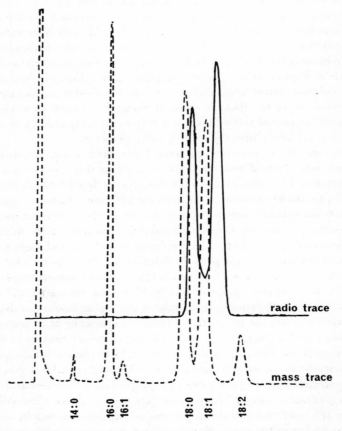

Fig. 10.2. Gas chromatographic separation of ^{14}C-labelled fatty acids with radioassay of components (trace obtained with Panax Radiogas Detector System).

loss of resolution of radioactivity peaks in comparison to the mass peaks is inevitable, even in the most efficient system, because of the dilution of the sample and the large volume of the apparatus through which it must pass after leaving the column. As well as the graphic record, the signal from the ratemeter can be passed into a scaler unit with a digital print-out so that the numbers of counts in predetermined time intervals are obtained directly.

When the background count is low, the number of counts recorded from a sample is proportional to the time the sample is in the detector cell which is in turn dependent on the gas flow rate and the volume of the detector cell. Higher counting efficiencies can, therefore, be obtained by reducing the flow of carrier gas and in instruments with a small detector cell, this improvement can be spectacular. For example, if the detector cell volume is 10 ml, a flow rate of 10 ml/min will give virtually complete recovery of radioactivity, whereas the more usual carrier gas flow rate of 50 ml/min will permit a counting efficiency of only 20 per cent.[521] In instruments of this type with small cells it is then advisable to use the lowest flow rate commensurate with the required degree of resolution in the gas chromatographic column, compensating for the increased retention time of components by raising the column temperature. Although it is always advisable to calibrate the instrument with precise amounts of $^{14}CO_2$ or $^3H_2$, it is also possible to remove doubts about the efficiency of the counting process by adding a internal standard of known specific activity to the sample. $[1-^{14}C]$-Pentadecanoic acid and $[1-^{14}C]$-heptadecanoic acid are used in this way in the analysis of $[^{14}C]$-labelled fatty acids in the author's laboratory.

The range of compounds that can be analysed by radio-gas chromatography in this manner depends on the temperature stability of the Viton seals in the connection to the stream splitter at the end of the column and the procedure has been used principally for the analysis of labelled fatty acids and compounds of similar volatility. Viton seals are not used in the system of Breckenridge and Kuksis[60] in which the metal part of a Kovar glass to metal seal is soldered to a stainless steel tube that is in turn connected to the stream splitter by means of a Swagelok fitting; $[^{14}C]$-triglycerides have been analysed with this system.

The principal disadvantage of continuous flow systems is that the compounds to be assayed must have comparatively high specific activities as the overall efficiency of the entire chromatographic and counting process may be low. There should be sufficient activity in a sample of less than 1 mg as the chromatographic columns are unlikely to take large amounts without loss of resolution. Radioactive decay is a random process and for a given number of counts (C) in a radioactive peak, the standard deviation (σ) is given by

$$\sigma = \sqrt{C}$$

It is therefore necessary to have 10,000 counts in a peak to obtain a 1 per cent standard deviation while 100 counts would give a 10 per cent deviation and it is not possible to extend the counting time and so improve the accuracy of the measurement as in liquid scintillation counting.

If need be, it is possible to by-pass the proportional counter and collect the $^{14}CO_2$ formed in the combustion unit in a scintillation vial containing 10 ml toluene–fluor and hyamine hydroxide in the ratio 15:8 (v/v). Fractions collected in this way can then be counted in a liquid scintillation counter with greater efficiency and an improved statistical accuracy.[541] It is claimed that better recoveries than are possible with conventional preparative GLC techniques are obtained in this way.

D. Location of ^{14}C in Aliphatic Chains

The portion of a radioactive lipid that contains the label can be ascertained simply by hydrolysing the lipid and isolating the various component parts, using the procedures described in earlier chapters, and assaying the radioactivity of each in a liquid scintillation counter. It may then be important to discover which carbon atoms in the aliphatic chains carry the ^{14}C-label. The reactions that are used to solve the problem are most easily performed with fatty acids and it is usually necessary to oxidise fatty alcohols or aldehydes to acids first. Before proceeding to the analysis, the individual fatty acids concerned must be isolated in the greatest degree of purity possible, using combinations of methods described in Chapter 5, so that traces of contaminants of high specific activity do not influence the results.

Fatty acids are more easily labelled with ^{14}C in the carboxyl group than in any other position by the available synthetic methods and, as a result, most ^{14}C-labelled fatty acids that are used in biological experi-

Reaction a.

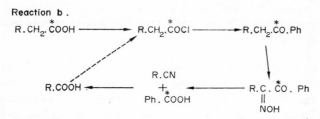

Fig. 10.3. Reaction a, Schmidt reaction for degradation of fatty acids. Reaction b, Stepwise degradation of radioactive fatty acids.

ments contain the label in this position. Methods of determining the activity of C-1 in fatty acids therefore have considerable importance. Of these, the Schmidt reaction, in which the fatty acid is reacted with hydrazoic acid (sodium azide and sulphuric acid) causing the carboxyl group to be released as carbon dioxide that can be recovered for liquid scintillation counting (reaction a, Fig. 10.3), is simplest and most suitable for the purpose. A number of semi-micro modifications of the procedure have been described for use with ^{14}C-labelled fatty acids and the following is based on that of Aronsson and Gürtler.[29] It is necessary to hydrogenate unsaturated fatty acids before carrying out the reaction.

"A micro-glass apparatus is required consisting of a flask and condenser on which is placed a pressure-equalising dropping funnel connected to an argon line which is led through the condenser to the bottom of the flask; there is a side-arm outlet from the top of the condenser, below the funnel, so that the gases pass into a suitable trap. The whole is set on top of a magnetic stirrer-heater. All the reagents should be of the highest possible purity. The ^{14}C-labelled fatty acid and a fatty acid carrier (40 mg in total) in dry benzene (2 ml) are added to the flask together with sodium azide (75 mg). 2-Methoxy-ethanol (2 ml) and ethanolamine (1 ml) are placed in the CO_2 collection vessel. The argon flow is set at 2 ml/min and concentrated sulphuric acid (0·5 ml) is introduced into the flask from the dropping funnel. The reaction is allowed to proceed at 55°C with stirring for 1 hr when the contents of the CO_2 trap are transferred to a scintillation vial with 10 ml of a toluene-based fluor (toluene, 268 ml; 2-methoxy-ethanol, 134 ml; 2,5-diphenyloxazole, 2 g) for counting. $^{14}CO_2$ recoveries of 99 per cent are possible but this should be checked with standards."

Unfortunately, methods of locating ^{14}C in other parts of a fatty acid chain tend to be less elegant and precise and often considerably more tedious than that described above for C-1. Sometimes information can be obtained by using basic structural features of the fatty acids; for example, unsaturated fatty acids can be oxidised at double bonds (by one of the methods described in Chapter 5) and the mono- and dibasic fragments formed can be separated and counted by radio-gas chromatography so that the distribution of radioactivity between the proximal and distal portions of the molecule is obtained. Polyunsaturated fatty acids are better subjected to partial reduction to obtain monoenoic acids (also discussed in Chapter 5) for structural analysis in this manner. A related procedure has been described by Stearns and Quackenbush[514] in which the double bonds of monoenoic fatty acids produced by partial reduction of polyunsaturated acids are reacted with a catalyst that causes the maximum possible isomerisation along the fatty acid chain; the isomerised fatty acids are then oxidised and the fragments subjected to radio-gas chromatography to determine the specific activities of each. The vigorous permanganate oxidation reaction that is used to locate methyl branches in aliphatic chains (see Chapter 5)[367, 378] can also be adapted to determining the radioactivity of specific carbon atoms in the chain of saturated fatty acids; the homologous series of fatty acids produced in the reaction must again be assayed by radio-gas chromatography. Although simple in concept, methods that produce large numbers of fragments suffer from the disadvantages that the fatty acids must initially have a very high activity so that sufficient remains in each of the fragments for the radio-gas chromatographic analyses to be meaningful and that the calculation of the activities of particular carbon atoms may be fraught with a number of errors if more than one carbon contains the label.

Methods of degrading fatty acids in a stepwise manner have been discussed in detail by Mead and Howton.[342] No simple method exists but the procedure of Dauben *et al.*[109] outlined in Fig. 10.3 (reaction b), though complex, avoids the use of strong oxidising agents so that over-reaction is unlikely to occur. The acid is converted to the acid chloride and thence to a phenylketone in a Friedel–Crafts condensation; the ketone is reacted with nitrous acid forming an α-oximinoketone that is cleaved by reaction with p-toluene-sulphonyl chloride and alkali to give benzoic acid and an aliphatic nitrile one carbon atom shorter than the original fatty acid. Both products can be recovered for radioassay and the nitrile hydrolysed to a fatty acid that can be put through the cycle again. Steinberg *et al.*[518] were able to obtain an overall yield of 65 per cent of the next lower homologue when the products were isolated by chromatographic procedures.

Determinations of the precise point of labelling of fatty acid metabolites isolated from biological experiments should undoubtedly be performed much more frequently than is the case at present, especially when the label was originally present in C-1, as it is well established that partial β-oxidation of fatty acids occurs to a considerable extent in mammalian tissues; labelled acetate is then released that can be assimilated into other fatty acids. Similarly, labelled acetate fed to animals or incubated with animal tissues *in vitro* may be used to resynthesise fatty acids that have been partially degraded. It is not always realised how prevalent these reactions are.

CHAPTER 11

A Summary

A. Introduction

The approach of the research worker to the analysis of the lipids in a given sample will depend partly on the amount of material in the sample, partly on the equipment and instrumentation available, but principally on the amount of information that is required. For example, if large numbers of samples must be analysed routinely, it may only be possible to perform the more basic compositional analyses on each one, while in other circumstances a detailed knowledge of the composition and structure of all the lipid components of a single sample may be required. Throughout this book it has been assumed that apparatus for gas–liquid and thin-layer chromatography will be available to the analyst.

In all analytical procedures, precautions must be taken to minimise the effects of autoxidation (see Chapter 3); in particular, lipids must be handled in an atmosphere of nitrogen wherever possible and anti-oxidants should be added to TLC sprays or to TLC solvent mixtures. Care should also be taken to prevent the introduction of contaminants into samples.

B. Extraction of Lipids from Tissues

Detailed extraction procedures are described in Chapter 2. The method of choice depends partly on the nature of the sample, e.g. plant tissues are extracted first with *iso*-propanol to prevent enzymatic degradation of lipids, and also on the amount of the sample. For reasons of economy in time and materials, simplified extraction procedures such as that of Bligh and Dyer[52] may be preferred for very large samples or for numbers of small samples although more exhaustive extraction procedures must be used when highly detailed analyses are intended. The analyst must also consider, when choosing a method of removing non-lipid contaminants from extracts, whether the gangliosides are required for analysis. The weight of the tissue and of the lipid obtained must be recorded and, in some circumstances, the amount of dry matter in the sample should be determined. This data can sometimes be obtained from a small representative aliquot of the sample.

C. Fatty Acid Compositions

Methods of determining the fatty acid compositions of lipid samples are discussed in Chapter 5. The methyl ester derivatives are first prepared (Chapter 4) and the method chosen will depend on the nature of the sample. If free fatty acids are present, for example, an acidic reagent such as methanolic hydrogen chloride must be used although the milder and more rapid alkaline transesterification reagents are to be preferred for glyceride-bound fatty acids (special procedures may be required for short-chain, amide-bound or unusual fatty acids). Infrared and ultra-violet spectroscopy may be of assistance in indicating the presence of some of the less common functional groups in fatty acid chains and ad-sorption chromatography (especially TLC) is of value in indicating whether these are polar in nature. Gas–liquid chromatography is the chief method of determining the fatty acid composition of lipids and analyses should preferably be performed on more than one type of polyester liquid phase. Components can be identified provisionally by their retention times relative to authentic standards, by their equivalent chain length values, on the basis of possible biosynthetic relationships or by their behaviour in other ancillary chromatographic techniques such as silver nitrate TLC. For unequivocal identification of a com-ponent, however, it is necessary to isolate it by some chromatographic procedure or combination of procedures and establish its structure by definitive chemical and/or spectroscopic techniques. The GLC quanti-fication method selected should be checked and calibrated regularly with standard mixtures of known composition similar in nature to the samples to be analysed. Results for each component fatty acid in a lipid class or mixture of lipids are generally expressed as "weight per cent" of the total but they must be converted to "moles per cent" for structural studies of lipids. In some circumstances, it is necessary to express results as "weight of each fatty acid per unit of tissue", but provided the weight per cent of lipid in the tissue is recorded, the methods of reporting data are interconvertible.

D. The Analysis of Simple Lipids

This topic is disscussed in detail in Chapter 6. Thin-layer chromatography is usually the method chosen for the analysis of simple lipids and it may be performed on the microgram to 100 mg scale. Layers of silica gel G 0·5 mm thick are preferred for larger amounts and those 0·25 mm thick for micro-amounts; hexane–diethyl ether–formic acid (80:20:2, by vol.) is by far the most widely used developing solvent. A few simple lipids can be identified by specific spray reagents (cholesterol and derivatives, free fatty acids or ester bonds) but normally authentic standards are run alongside the samples under investigation. With very small amounts of material or with large numbers of similar samples, fluorometry or charring followed by photodensitometry are generally the methods of choice for lipid quantification; this can be performed while the samples are on the plate or after they have been eluted from the adsorbent. With large amounts of lipid (i.e. 2 mg or more), specific chemical techniques can be applied to quantifying individual components or gas chromatography of the fatty acid constituents of each lipid class in the presence of a known amount of a suitable internal standard may be used so that the fatty acid composition and the amount of each fraction are determined in the same analysis. When larger amounts of particular simple lipids are required for structural analyses say, they can be isolated by column chromatography on silicic acid or Florisil although, because of the difficulty of monitoring columns, it may often be simpler to fractionate samples by preparative TLC on several identical TLC plates with pooling of corresponding fractions.

Alkyl diglycerides and neutral plasmalogens are sometimes found with triglycerides in lipid samples and can be separated from each other with care by thin layer chromatography. They can be determined by hydrogenolysis with lithium aluminium hydride with TLC separation of the products. Aldehydes, liberated from plasmalogens by acidic conditions, may be chromatographed as such or in the form of more stable derivatives on similar GLC columns as are used for the analysis of methyl esters of fatty acids.

Results obtained by the above quantification procedures are most often expressed for each component in terms of "weight per cent of the total lipid" although sometimes "weight per unit of tissue" is preferred. Provided that the total weight of lipid per unit of tissue is recorded, however, either method is suitable.

E. The Analysis of Complex Lipids

Procedures for the analysis of complex lipids are discussed in Chapter 7. Complex lipids can be separated from simple lipids for analysis by rapid column chromatographic procedures on virtually any scale and if glycolipids are also major components of the sample, they can also be isolated at this stage. With small amounts of material the complex lipids can be obtained by preparative TLC using the system described in the above section.

The more common phospholipids from animal tissues can be separated by TLC on silica gel H layers in a single dimension (on up to the 10 mg scale) with chloroform–methanol–acetic acid–water (25:15:4:2, by vol.) as developing solvent.[491] Components can be identified by their chromatographic behaviour relative to authentic standards or by means of specific spray reagents. Lipids can be determined by phosphorus analysis or by GLC of their constituent fatty acids with a known amount of an internal standard so that the fatty acid composition and amount of each lipid class are determined in the same analysis. Other TLC systems in one direction are available for the analysis of the acidic lipids that are generally minor components but two-dimensional TLC systems are a valuable alternative as they permit the separation of complex lipids into many more fractions on one TLC plate than is possible with one-dimensional systems. Related procedures are available for plant and bacterial lipids.

Glycosphingolipids can be separated from phosphatides by several chromatographic or chemical procedures and can be fractionated into single components differing in the number and nature of the hexose units by one-dimensional TLC and estimated by determining the amounts of the nitrogenous bases by chemical procedures. Chroma-

Lipid Analysis

tographic methods are also available for determining the fatty acid composition, the long-chain base composition and the hexose composition of individual glycosphingolipids.

When detailed analyses of all the minor components are required, more material must be used and it is necessary to convert all the components to the same salt form before commencing the separation process. Combinations of techniques such as DEAE or TEAE cellulose column chromatography in conjunction with preparative TLC or other column procedures must then be used to separate all the lipid components.

Alkyl- and alkenyl- forms of individual phosphatides cannot be separated from the diacyl-form of the phospholipid in the natural state but they can be determined by the hydrogenolysis procedure used for the analogous simple lipids. If the polar phosphorus group is removed or modified chemically, however, separation of the various forms of the phosphatide is sometimes possible.

Results of analyses are expressed in a variety of ways but generally reflect the method used for quantification. For example, "moles per cent phosphorus" or "weight of phospholipid per unit of tissue" (using the conversion factor of 25) are often used for each component when phosphorus analysis has been performed although "moles per cent phospholipid" can also be calculated from this data. "Moles per cent phospholipid" is normally the method of choice when GLC internal standard procedures are used for quantification. "Moles per cent of the total glycolipids" is the simplest method of expressing the results of glycosphingolipid analyses as these depend generally on determinations of the molar amounts of the nitrogenous bases. For the sake of uniformity, the author would prefer to see all results converted to "moles per cent of the total complex lipids".

F. Structural Analysis of Lipids

Methods of separating lipids into molecular species are described in Chapter 8 and for determining the positional distribution of fatty acids in lipids in Chapter 9. Lipids can be separated into simpler molecular species according to the combined properties of all the fatty acid constituents by silver nitrate and/or reversed phase TLC and often by high temperature GLC. Combinations of two of these procedures should preferably be used wherever this is feasible. It is usually necessary to convert phospholipids to less polar derivatives prior to analysis in this manner, either by removing the phosphorus group entirely by means of phospholipase C hydrolysis, when the resulting diglyceride or a derivative thereof can be subjected to high temperature GLC in addition to the TLC separations, or by rendering the phosphate group less polar by enzymic and chemical means, when compounds labelled with ^{32}P can be studied. GLC of the component fatty acids with an added internal standard or related procedures are generally the methods chosen for determining molecular species separated in this way.

Complicated stereospecific analysis procedures have been developed for determining the distribution of fatty acids in positions 1, 2 and 3 of L-glycerol in triglycerides but the composition of position 2 alone is obtained relatively easily by means of hydrolysis with pancreatic lipase. Phospholipase A_2 can be used to establish the distribution of fatty acids between positions 1 and 2 of glycerophosphatides.

In all lipid structural studies, results for fatty acid compositions and molecular species should always be expressed in terms of "moles per cent" of the total.

APPENDIX A

Commercial Sources of Lipid Standards

Analabs Inc., P.O. Box 501, North Haven, Con. (06473), U.S.A. U.K. Distributor—Auriema Ltd., 442 Bath Road, Slough, Bucks.

Applied Science Laboratories, Inc., P.O. Box 440, State College, Pa. (16801), U.S.A. U.K. Distributor—Field Instrument Co., Tetrapack House, Orchard Road, Richmond, Surrey.

B.D.H. Chemicals Ltd., Poole, Dorset, U.K.

Hormel Institute, University of Minnesota, 801 16th Ave N.E., Austin, Minn. (55912), U.S.A.

Nu-Chek-Prep, Inc., P.O. Box 172, Elysian, Minn. (56028), U.S.A.

P–L Biochemicals Inc., 1037 West McKinley Ave., Milwaukee, Wisc., U.S.A.

Serdary Research Laboratories, 1643 Kathryn Drive, London, Ontario, Canada.

Sigma Chemical Co., P.O. Box 14508, Saint Louis, Missouri (63178), U.S.A. U.K. Distributor—Sigma London Chemical Co. Ltd., Norbiton Station Yard, Kingston-upon-Thames, Surrey.

Supelco Inc., Bellefonte, Pa. (16823), U.S.A. U.K. Distributor—Field Instrument Co., Tetrapack House, Orchard Road, Richmond, Surrey.

APPENDIX B

Sources of Information

Journals

The following journals are devoted entirely to papers on the chemistry, biochemistry and analysis of lipids.

Biochimica Biophysica Acta—Lipids and Lipid Metabolism Section
Chemistry and Physics of Lipids
Journal of the American Oil Chemists' Society
Journal of Lipid Research
Lipids

In addition many other journals publish papers on the subject including—*Archives of Biochemistry and Biophysics, Biochemical Journal, Biochemistry, Canadian Journal of Biochemistry, Comparative Biochemistry and Physiology, Hoppe-Seyler's Zeitschrift für Physiologische Chemie, Journal of Biochemistry (Tokyo), Journal of Biological Chemistry, Journal of Chromatography.*

Reviews

The following review volumes appear annually and are devoted entirely to various aspects of the study of lipids:

Progress in the Chemistry of Fats and other Lipids (edited by R. T. Holman, Pergamon Press, London). Thirteen volumes to date.
Advances in Lipid Research (edited by R. Paoletti and D. Kritchevsky, Academic Press, New York). Ten volumes to date.
Topics in Lipid Chemistry (edited by F. D. Gunstone, Logos Press and Paul Elek Books). Three volumes to date.

Introductory Books

The following books can be recommended for the newcomer to the subject.

Introduction to Lipids, by D. Chapman (McGraw-Hill, London, 1969).

An Introduction to the Chemistry and Biochemistry of Fatty Acids and their Glycerides, by F. D. Gunstone (Chapman and Hall Ltd., London, 1967).

Lipid Biochemistry: An Introduction, by M. I. Gurr and A. T. James (Chapman and Hall Ltd., London, 1971).

Physiological Chemistry of Lipids in Mammals, by E. J. Masoro (Saunders, Philadelphia, 1968).

More Advanced Texts

Comprehensive Biochemistry, Vol. 18. *Lipid Metabolism*, edited by M. Florkin and E. H. Stotz (Elsevier, Amsterdam, 1970).

Methods in Enzymology, Vol. 14. *Lipids*, edited by J. M. Lowenstein (Academic Press, New York, 1970).

Lipid Chromatographic Analysis, Vol. 1, edited by G. V. Marinetti (Edward Arnold Ltd., London, 1967).

Plant Lipid Biochemistry, by C. Hitchcock and B. W. Nichols (Academic Press, London, 1971).

Lipids and Lipidoses, edited by G. Schettler (Springer-Verlag, Berlin, 1967).

Structural and Functional Aspects of Lipoproteins in Living Systems, edited by E. Tria and A. M. Scanu (Academic Press, London, 1969).

Lipid Metabolism, edited by S. J. Wakil (Academic Press, New York, 1970).

Analysis of Triglycerides, by C. Litchfield (Academic Press, New York, 1972).

L

References

1. AASEN, A. J., HOFSTETTER, H. H., IYENGAR, B. T. R. and HOLMAN, R. T., *Lipids*, **6**, 502 (1971).
2. ABRAHAMSSON, S., STÄLLBERG-STENHAGEN, S. and STENHAGEN, E., *Progress in the Chemistry of Fats and other Lipids*, Vol. 7, p. 1 (1964) (edited by R. T. Holman, Pergamon Press, London).
3. ACKMAN, R. G., *J. Amer. Oil Chem. Soc.*, **40**, 558 (1963).
4. ACKMAN, R. G., *J. Gas Chromatogr.*, **2**, 173 (1964).
5. ACKMAN, R. G., *J. Chromatogr.*, **28**, 225 (1967).
6. ACKMAN, R. G., *Lipids*, **2**, 502 (1967).
7. ACKMAN, R. G., *J. Chromatogr.*, **34**, 165 (1968).
8. ACKMAN, R. G., *Methods in Enzymology*, Vol. XIV, p. 329 (1969) (edited by J. M. Lowenstein, Academic Press, New York).
9. ACKMAN, R. G., *Progress in the Chemistry of Fats and other Lipids*, Vol. 12, p. 165 (1972) (edited by R. T. Holman, Pergamon Press, London).
10. ACKMAN, R. G. and BURGHER, R. D., *J. Amer. Oil Chem. Soc.*, **42**, 38 (1965).
11. ACKMAN, R. G. and HANSEN, R. P., *Lipids*, **2**, 357 (1967).
12. ACKMAN, R. G. and SIPOS, J. C., *J. Chromatogr.*, **16**, 298 (1964).
13. ACKMAN, R. G., SIPOS, J. C. and JANGAARD, P. M., *Lipids*, **2**, 251 (1967).
14. AGRANOFF, B. W., BRADLEY, R. M. and BRADY, R. O., *J. Biol. Chem.*, **233**, 1077 (1958).
15. ÅKESSON, B., *Eur. J. Biochem.*, **9**, 463 (1969).
16. AKINO, T. and SHIMOJO, T., *Biochim. Biophys. Acta*, **210**, 343 (1970).
17. ALBRO, P. W. and DITTMER, J. C., *J. Chromatogr.*, **38**, 230 (1968).
18. ALLEN, R. R., *J. Amer. Oil Chem. Soc.*, **46**, 552 (1969).
19. AMENTA, J. S., *J. Lipid Res.*, **5**, 270 (1964).
20. AMES, B. N. and DUBIN, D. T., *J. Biol. Chem.*, **235**, 769 (1960).
21. ANDERSON, R. E., BOTTINO, N. R. and REISER, R., *Lipids*, **2**, 440 (1967).
22. ANDERSON, R. E., GARRETT, R. D., BLANK, M. L. and SNYDER, F., *Lipids*, **4**, 327 (1969).
23. ANDERSON, R. L. and HOLLENBACH, E. J., *J. Lipid Res.*, **6**, 577 (1965).
24. ANDO, N., ANDO, S. and YAMAKOWA, T., *J. Biochem. (Tokyo)*, **70**, 341 (1971).
25. ANTHONY, G. M., BROOKS, C. J. W., MACLEAN, I. and SANGSTER, I., *J. Chromatogr. Sci.*, **7**, 623 (1969).
26. Applied Science Laboratories Inc., *Gas-Chrom Newsletter*, Vol. 11, No. 4 (1970).
27. ARCUS, A. C. and DUNCKLEY, G. G., *J. Chromatogr.*, **5**, 272 (1961).
28. ARGOUDELIS, C. J. and PERKINS, E. G., *Lipids*, **3**, 379 (1968).
29. ARONSSON, P. and GÜRTLER, J., *Biochim. Biophys. Acta*, **248**, 21 (1971).
30. ARUNGA, R. O. and MORRISON, W. R., *Lipids*, **6**, 768 (1971).
31. ARVIDSON, G. A. E., *J. Lipid Res.*, **6**, 574 (1965).

32. ARVIDSON, G. A. E., *J. Lipid Res.*, **8**, 155 (1967).
33. ARVIDSON, G. A. E., *Eur. J. Biochem.*, **4**, 478 (1968).
34. AUDIER, H., BORY, S., FÉTIZON, M., LONGEVIALLE, P. and TOUBIANA, R., *Bull. Soc. Chim. France*, 3034 (1964).
35. AULING, G., HEINZ, E. and TULLOCH, A. P., *Hoppe-Seyler's Z. Physiol. Chem.*, **352**, 905 (1971).
36. BAER, E., *Lipids*, **3**, 384 (1968).
37. BAGBY, M. O., SMITH, C. R. and WOLFF, I. A., *J. Org. Chem.*, **30**, 4227 (1965).
38. BARFORD, R. A., LUDDY, F. E. and MAGIDMAN, P., *Lipids*, **1**, 287 (1966).
39. BARRETT, C. B., DALLAS, M. S. J. and PADLEY, F. B., *J. Amer. Oil Chem. Soc.*, **40**, 580 (1963).
40. BARTLETT, G. R., *J. Biol. Chem.*, **234**, 466 (1959).
41. BARTLET, J. C. and IVERSON, J. L., *J. Assoc. Off. Anal. Chem.*, **49**, 21 (1966).
42. BAUMANN, W. J. and MANGOLD, H. K., *J. Org. Chem.*, **29**, 3055 (1964).
43. BAUMANN, W. J. and ULSHÖFER, H. W., *Chem. Phys. Lipids*, **2**, 114 (1968).
44. BERGELSON, L. D., *Progress in the Chemistry of Fats and other Lipids*, Vol. 10, p. 239 (1970) (edited by R. T. Holman, Pergamon Press, London).
45. BERGER, H. and HANAHAN, D. J., *Biochim. Biophys. Acta*, **231**, 584 (1971).
46. BIERL, B. A., BEROZA, M. and ALDRIDGE, M. H., *Anal. Chem.*, **43**, 636 (1971).
47. BIRKOFER, L., RITTER, A. and BENTZ, F., *Chem. Ber.*, **97**, 2119 (1964).
48. BISCHEL, M. D. and AUSTIN, J. H., *Biochim. Biophys. Acta*, **70**, 598 (1963).
49. BISHOP, C., GLASCOCK, R. F., NEWELL, E. M. and WELCH, V. A., *J. Lipid Res.*, **12**, 777 (1971).
50. BLANK, M. L., SCHMIT, J. A. and PRIVETT, O. S., *J. Amer. Oil Chem. Soc.*, **41**, 371 (1964).
51. BLANK, M. L. and SNYDER, F., *Biochemistry*, **9**, 5034 (1970).
52. BLIGH, E. G. and DYER, W. J., *Can. J. Biochem. Physiol.*, **37**, 911 (1959).
53. BOLDINGH, J., *Rec. Trav. Chim.*, **69**, 247 (1950).
54. BORKA, L. and PRIVETT, O. S., *Lipids*, **1**, 104 (1966).
55. BOROWSKI, E. and ZIMINSKI, T., *J. Chromatogr.*, **23**, 480 (1966).
56. BOTTINO, N. R., *J. Lipid Res.*, **12**, 24 (1971).
57. BOTTINO, N. R., VANDENBURG, G. A. and REISER, R., *Lipids*, **2**, 489 (1967).
58. BOYNE, A. W. and DUNCAN, W. R. H., *J. Lipid Res.*, **11**, 293 (1970).
59. BRANDT, A. E. and LANDS, W. E. M., *Lipids*, **3**, 178 (1968).
60. BRECKENRIDGE, W. C. and KUKSIS, A., *Lipids*, **5**, 342 (1970).
61. BRIAN, B. L. and GARDNER, E. W., *Appl. Microbiol.*, **16**, 549 (1968).
62. BROCKERHOFF, H., *J. Lipid Res.*, **6**, 10 (1965).
63. BROCKERHOFF, H., *Arch. Biochem. Biophys.*, **110**, 586 (1965).
64. BROCKERHOFF, H., *Lipids*, **1**, 162 (1966).
65. BROCKERHOFF, H., *J. Lipid Res.*, **8**, 167 (1967).
66. BROCKERHOFF, H., *Lipids*, **6**, 942 (1971).
67. BROOKS, C. J. W. and MACLEAN, I., *J. Chromatogr. Sci.*, **9**, 18 (1971).
68. BROWN, H. C., SIVASANKARAN, K. and BROWN, C. A., *J. Org. Chem.*, **28**, 214 (1963).
69. BROWN, J. B. and KOLB, D. K., *Progress in the Chemistry of Fats and other Lipids*, Vol. 3, p. 57 (1955) (edited by R. T. Holman, Pergamon Press, London).
70. BRUNDISH, D. E., SHAW, N. and BADDILEY, J., *Biochem. J.*, **97**, 158 (1965).
71. BU'LOCK, J. D. and SMITH, G. N., *J. Chem. Soc.*, (**c**), 322 (1967).
72. CAPELLA, P. and ZORZUT, C. M., *Anal. Chem.*, **40**, 1458 (1968).
73. CARROLL, K. K., *J. Lipid Res.*, **2**, 135 (1961).

74. CARROLL, K. K., *J. Amer. Oil Chem. Soc.*, **40,** 413 (1963).
75. CARROLL, K. K. and SERDAREVICH, B., *Lipid Chromatographic Analysis*, Vol. 1, p. 205 (1967) (edited by G. V. Marinetti, Edward Arnold Ltd., London).
76. CARTER, H. E., ROTHFUS, J. A. and GIGG, R., *J. Lipid Res.*, **2,** 228 (1961).
77. CARTER, H. E., SMITH, D. B. and JONES, D. N., *J. Biol. Chem.*, **232,** 681 (1958).
78. CARTER, H. E., STROBACH, D. R. and HAWTHORNE, J. N., *Biochemistry*, **8,** 383 (1969).
79. CARTER, H. E. and WEBER, E. J., *Lipids*, **1,** 16 (1966).
80. CASPARRINI, G., HORNING, E. C. and HORNING, M. G., *Chem. Phys. Lipids*, **3,** 1 (1969).
81. CASTER, W. O., AHN, P. and POGUE, R., *Chem. Phys. Lipids*, **1,** 393 (1967).
82. CHAPMAN, D., *J. Amer. Oil Chem. Soc.*, **42,** 353 (1965).
83. CHAPMAN, D., *The Structure of Lipids by Spectroscopic and X-ray Techniques*, (1965) (Methuen and Co. Ltd., London).
84. CHERNICK, S. S., *Methods in Enzymology*, Vol. XIV, p. 627 (1969) (edited by J. M. Lowenstein, Academic Press, New York).
85. CHRISTIANSEN, K., MAHADEVAN, V., VISWANATHAN, C. V. and HOLMAN, R. T., *Lipids*, **4,** 421 (1969).
86. CHRISTIE, W. W., *J. Chromatogr.*, **34,** 405 (1968).
87. CHRISTIE, W. W., *J. Chromatogr.*, **37,** 27 (1968).
88. CHRISTIE, W. W., *Biochim. Biophys. Acta*, **187,** 1 (1969).
89. CHRISTIE, W. W., *Topics in Lipid Chemistry*, Vol. 1, p. 1 (1970) (edited by F. D. Gunstone, Logos Press, London).
90. CHRISTIE, W. W., *Analyst (London)*, **97,** 221 (1972).
91. CHRISTIE, W. W., *Topics in Lipid Chemistry*, Vol. 3, p. 171 (1972) (edited by F. D. Gunstone, Logos Press, London).
92. CHRISTIE, W. W., GUNSTONE, F. D., ISMAIL, I. A. and WADE, L., *Chem. Phys. Lipids*, **2,** 196 (1968).
93. CHRISTIE, W. W., GUNSTONE, F. D. and PRENTICE, H. G., *J. Chem. Soc.*, 5768 (1963).
94. CHRISTIE, W. W., GUNSTONE, F. D., PRENTICE, H. G. and SEN GUPTA, S. C., *J. Chem. Soc.*, 5833 (1964).
95. CHRISTIE, W. W. and HOLMAN, R. T., *Chem. Phys. Lipids*, **1,** 407 (1967).
96. CHRISTIE, W. W. and MOORE, J. H., *Biochim. Biophys. Acta*, **176,** 445 (1969).
97. CHRISTIE, W. W. and MOORE, J. H., *Biochim. Biophys. Acta*, **210,** 46 (1970).
98. CHRISTIE, W. W. and MOORE, J. H., *J. Sci. Food Agric.*, **22,** 120 (1971).
99. CHRISTIE, W. W., NOBLE, R. C. and MOORE, J. H., *Analyst (London)*, **95,** 940 (1970).
100. CHRISTIE, W. W., REBELLO, D. and HOLMAN, R. T., *Lipids*, **4,** 229 (1969).
101. CHRISTOPHERSON, S. W. and GLASS, R. L., *J. Dairy Sci.*, **52,** 1289 (1969).
102. CHUNG, A. E. and GOLDFINE, H., *Nature*, **206,** 1253 (1965).
103. CLAYTON, T. A., MACMURRAY, T. A. and MORRISON, W. R., *J. Chromatogr.*, **47,** 277 (1970).
104. CONACHER, H. B. S., GUNSTONE, F. D., HORNBY, G. M. and PADLEY, F. B., *Lipids*, **5,** 434 (1970).
105. CARTONI, G., LIBERTI, A. and RUGGIERI, G., *Riv. Ital. Sostanze Grasse*, **40,** 482 (1963).
106. COWARD, R. F. and SMITH, P., *J. Chromatogr.*, **45,** 230 (1969).
107. CRAWFORD, R. V. and HILDITCH, T. P., *J. Sci. Food Agric.*, **1,** 230 (1950).

312   References

108. CROCKEN, B. J. and NYC, J. F., *J. Biol. Chem.*, **239**, 1727 (1964).
109. DAUBEN, W. G., HOERGER, E. and PETERSON, J. W., *J. Amer. Chem. Soc.*, **75**, 2347 (1953).
110. All references to this number deleted.
111. DAWSON, R. M. C., *Lipid Chromatographic Analysis*, Vol. 1, p. 163 (1967) (edited by G. V. Marinetti, Edward Arnold Ltd., London).
112. DAWSON, R. M. C. and EICHBERG, J., *Biochem. J.*, **96**, 634 (1965).
113. DAWSON, R. M. C. and KEMP, P., *Biochem. J.*, **105**, 837 (1967).
114. DE BOER, TH. J. and BACKER, H. J., *Organic Syntheses*, Coll. Vol. 4, p. 250 (1963) (edited by N. Rabjohn, John Wiley and Sons Inc., London).
115. DE HAAS, G. H. and VAN DEENEN, L. L. M., *Biochim. Biophys. Acta*, **106**, 315 (1965).
116. DHOPESHWARKAR, G. A. and MEAD, J. F., *Proc. Soc. Exp. Biol. Med.* **109**, 425 (1962).
117. DINH-NGUYÊN, N., RYHAGE, R. and STÄLLBERG-STENHAGEN, S., *Arkiv. Kemi*, **15**, 433 (1960).
118. DIRINGER, H., *Hoppe-Seyler's Z. Physiol. Chem.*, **353**, 39 (1972).
119. DITTMER, J. C., FEMINELLA, J. L. and HANAHAN, D. J., *J. Biol. Chem.*, **233**, 862 (1958).
120. DITTMER, J. C. and LESTER, R. L., *J. Lipid Res.*, **5**, 126 (1964).
121. DITTMER, J. C. and WELLS, M. A., *Methods in Enzymology*, Vol. XIV, p. 482 (1969) (edited by J. M. Lowenstein, Academic Press, New York).
122. DOLEV, A., ROHWEDDER, W. K. and DUTTON, H. J., *Lipids*, **1**, 231 (1966).
123. DOWNING, D. T. and GREENE, R. S., *Lipids*, **3**, 96 (1968).
124. DUDZINSKI, A. E., *J. Chromatogr.*, **31**, 560 (1967).
125. DUTTON, H. J., *J. Amer. Oil Chem. Soc.*, **38**, 631 (1961).
126. DYATLOVITSKAYA, E. V., VOLKOVA, V. E. and BERGELSON, L. D., *Bull. Acad. Sci. USSR, Div. Chem. Sci.*, 946 (1966).
127. EGLINTON, G. and HAMILTON, R. J., *Science*, **156**, 1322 (1967).
128. EMKEN, E. A., *Lipids*, **6**, 686 (1971).
129. EMKEN, E. A., *Lipids*, **7**, 459 (1972).
130. EMKEN, E. A., SCHOLFIELD, C. R. and DUTTON, H. J., *J. Amer. Oil Chem. Soc.*, **41**, 388 (1964).
131. ENTRESSANGLES, B., SARI, H. and DESNUELLE, P., *Biochim. Biophys. Acta*, **125**, 597 (1966).
132. ETTRE, L. S., *Open Tubular Columns*, 1965 (Plenum Press, New York).
133. ETTRE, L. S., PURCELL, J. E. and NOREM, S. D., *J. Gas Chromatogr.*, **31**, 181 (1965).
134. FANG, M. and MARINETTI, G. V., *Methods in Enzymology*, Vol. XIV, p. 598 (1969) (edited by J. M. Lowenstein, Academic Press, New York).
135. FERRELL, W. J., RADLOFF, J. F. and JACKIW, A. B., *Lipids*, **4**, 278 (1969).
136. FERRELL, W. J. and YAO, K-C., *J. Lipid Res.*, **13**, 23 (1972).
137. FEWSTER, M. E., BURNS, B. J. and MEAD, J. F., *J. Chromatogr.*, **43**, 120 (1969).
138. FOLCH, J., LEES, M. and STANLEY, G. H. S., *J. Biol. Chem.*, **226**, 497 (1957).
139. FREEMAN, C. P. and WEST, D., *J. Lipid Res.*, **7**, 324 (1966).
140. FREEMAN, N. K., *J. Amer. Oil Chem. Soc.*, **45**, 798 (1968).
141. FREEMAN, N. K., LINDGREN, F. T., NG, Y. C. and NICHOLS, A. V., *J. Biol. Chem.*, **227**, 449 (1957).
142. FROST, D. J. and BARZILAY, J., *Anal. Chem.*, **43**, 1316 (1971).
143. FULK, W. K. and SHORB, M. S., *J. Lipid Res.*, **11**, 276 (1970).

144. FUNAKASI, H. and GILBERTSON, J. R., *J. Lipid Res.*, **9**, 766 (1968).
145. GALANOS, D. S. and KAPOULAS, V. M., *J. Lipid Res.*, **3**, 134 (1962).
146. GALLAI-HATCHARD, J. J. and GRAY, G. M., *Biochim. Biophys. Acta*, **116**, 532 (1966).
147. GARDNER, H. W., *J. Lipid Res.*, **9**, 139 (1968).
148. GATT, S., *Methods in Enzymology*, Vol. XIV, p. 152 (1969) (edited by J. M. Lowenstein, Academic Press, U.S.A.).
149. All references to this number deleted.
150. GAVER, R. C. and SWEELEY, C. C., *J. Amer. Oil Chem. Soc.*, **42**, 294 (1965).
151. GAVER, R. C. and SWEELEY, C. C., *J. Amer. Oil Chem. Soc.*, **88**, 3643 (1966).
152. GEYER, K. G. and GOODMAN, H. M., *Proc. Soc. Exp. Biol. Med.*, **133**, 404 (1970).
153. GILBERTSON, J. R., GARLICH, H. H. and GELMAN, R. A., *J. Lipid Res.*, **11**, 201 (1970).
154. GILLILAND, K. M. and MOSCATELLI, E. A., *Biochim. Biophys. Acta*, **187**, 221 (1969)
155. GLASS, C. A. and DUTTON, H. J., *Anal. Chem.*, **36**, 2401 (1964).
156. GONZALEZ-SASTRE, F. and FOLCH-PI, J., *J. Lipid Res.*, **9**, 532 (1968).
157. GOTTFRIED, E. L. and RAPPORT, M. M., *J. Biol. Chem.*, **237**, 329 (1962).
158. GRAY, G. M., *Nature*, **207**, 505 (1965).
159. GRAY, G. M., *Lipid Chromatographic Analysis*, Vol. 1, p. 401 (1967) (edited by G. V. Marinetti, Edward Arnold Ltd., London).
160. GRAY, G. M., *Biochim. Biophys. Acta*, **144**, 511 (1967).
161. GRAY, G. M., *Biochim. Biophys. Acta*, **144**, 519 (1967).
162. GREEN, T., HOWITT, F. O. and PRESTON, R., *Chem. Ind. (London)*, 591 (1955).
163. GUNSTONE, F. D. and INGLIS, R. P., *Topics in Lipid Chemistry*, Vol. 2, p. 287 (1971) (edited by F. D. Gunstone, Logos Press, London).
164. GUNSTONE, F. D. and ISMAIL, I. A., *Chem. Phys. Lipids*, **1**, 337 (1967).
165. GUNSTONE, F. D., ISMAIL, I. A. and LIE KEN JIE, M., *Chem. Phys. Lipids*, **1**, 376 (1967).
166. GUNSTONE, F. D. and JACOBSBERG, F. R., *Chem. Phys. Lipids*, **9**, 26 (1972).
167. GUNSTONE, F. D., KILCAST, D., POWELL, R. G. and TAYLOR, G. H., *Chem. Commun.*, 295 (1967).
168. GUNSTONE, F. D. and LIE KEN JIE, M., *Chem. Phys. Lipids*, **4**, 131 (1970).
169. GUNSTONE, F. D., LIE KEN JIE, M. and WALL, R. T., *Chem. Phys. Lipids*, **3**, 297 (1969).
170. GUNSTONE, F. D. and MORRIS, L. J., *J. Chem. Soc.*, 2127 (1959).
171. GUNSTONE, F. D. and PADLEY, F. B., *J. Amer. Oil Chem. Soc.*, **42**, 957 (1965).
172. GUNSTONE, F. D., PADLEY, F. B. and QURESHI, M. I., *Chem. Ind. (London)*, 483 (1964).
173. GUNSTONE, F. D. and SUBBARAO, R., *Chem. Ind. (London)*, 461 (1966).
174. HAAHTI, E. and NIKKARI, T., *Acta Chem. Scand.*, **17**, 536 (1963).
175. HAAHTI, E., NIKKARI, T. and JUVA, K., *Acta Chem. Scand.*, **17**, 538 (1963).
176. HADORN, H. and ZUERCHER, K., *Mitt. Lebensmittelunters. Hyg.*, **58**, 236 (1967).
177. HAINES, T. H., *Progress in the Chemistry of Fats and other Lipids*, Vol. 11, p. 297 (1971) (edited by R. T. Holman, Pergamon Press, London).
178. HAJRA, A. K. and RADIN, N. S., *J. Lipid Res.*, **3**, 131 (1962).
179. HAKOMORI, S. and STRYCHARZ, G. D., *Biochemistry*, **7**, 1279 (1968).
180. HALLGREN, B. and LARSSON, S., *J. Lipid Res.*, **3**, 31 (1962).
181. HALLGREN, B., RYHAGE, R. and STENHAGEN, E., *Acta Chem. Scand.*, **13**, 845 (1959).
182. HAMILTON, R. J. and YAQUB RAIE, M., *Chem. Ind. (London)*, 1228 (1971).

183. HAMMARSTRÖM, S., *J. Lipid Res.*, **11,** 175 (1970).
184. HAMMERSTRÖM, S. and SAMUELSSON, B., *J. Biol. Chem.*, **247,** 1001 (1972).
185. HAMMARSTRÖM, S., SAMUELSSON, B. and SAMUELSSON, K., *J. Lipid Res.*, **11,** 150 (1970).
186. HAMMOND, E. G., *Lipids*, **4,** 246 (1969).
187. HANAHAN, D. J., EKHOLM, J. and JACKSON, C. M., *Biochemistry*, **2,** 630 (1963).
188. HANAHAN, D. J. and OLLEY, J. N., *J. Biol. Chem.*, **321,** 813 (1958).
189. HANAHAN, D. J., TURNER, M. B. and JAYKO, M. E., *J. Biol. Chem.*, **192,** 623 (1951).
190. HANSEN, R. P., *Chem. Ind. (London)*, 1640 (1967).
191. HANSEN, R. P. and SMITH, J. F., *Lipids*, **1,** 316 (1966).
192. HARLOW, R. D., LITCHFIELD, C. and REISER, R., *Lipids*, **1,** 216 (1966).
193. HAVERKATE, F. and VAN DEENEN, L. L. M., *Biochim. Biophys. Acta*, **106,** 78 (1965).
194. HAWTHORNE, J. N. and KEMP, P., *Advances in Lipid Research*, Vol. 2, p. 127 (1964) (edited by R. Paoletti and D. Kritchevsky, Academic Press, London).
195. HAY, J. D. and MORRISON, W. R., *Biochim. Biophys. Acta*, **202,** 237 (1970).
196. HEINZ, E., *Biochim. Biophys. Acta*, **144,** 333 (1967).
197. HEINZ, E. and TULLOCH, A. P., *Hoppe-Seyler's Z. Physiol. Chem.*, **350,** 493 (1969).
198. HELMY, F. M. and HACK, M. H., *Lipids*, **1,** 279 (1966).
199. HENLY, R. S., *J. Amer. Oil Chem. Soc.*, **42,** 673 (1965).
200. HENLY, R. S. and ROYER, D. J., *Methods in Enzymology*, Vol. XIV, p. 450 (1969) (edited by J. M. Lowenstein, Academic Press, New York).
201. HEYNEMAN, R. A., BERNARD, D. M. and VERCANTEREN, R. E., *J. Chromatogr.*, **68,** 285 (1972).
202. HILL, E. E., HUSBANDS, D. R. and LANDS, W. E. M., *J. Biol. Chem.*, **243,** 4440 (1968).
203. HIRSCH, J., *Coloq. Intern. Centre Nat. Rech. Sci., Paris*, **99,** 11 (1961).
204. HIRSCH, J. and AHRENS, E. H., *J. Biol. Chem.*, **233,** 311 (1958).
205. HITCHCOCK, C. and NICHOLS, B. W., *Plant Lipid Biochemistry*, Chapter 2, p. 43 (1971) (Academic Press, London).
206. HITES, R. A., *Anal. Chem.*, **42,** 1736 (1970).
207. HOEVET, S. P., VISWANATHAN, C. V. and LUNDBERG, W. O., *J. Chromatogr.*, **34,** 195 (1968).
208. HOFSTETTER, H. H., SEN, N. and HOLMAN, R. T., *J. Amer. Oil Chem. Soc.*, **42,** 537 (1965).
209. HOLLA, K. S. and CORNWELL, D. G., *J. Lipid Res.*, **6,** 322 (1965).
210. HOLLA, K. S., HORROCKS, L. A. and CORNWELL, D. G., *J. Lipid Res.*, **5,** 263 (1964).
211. HOLLOWAY, P. J. and CHALLEN, S. B., *J. Chromatogr.*, **25,** 336 (1966).
212. HOLMAN, R. T., *Methods Biochem. Anal.*, **2,** 113 (1955).
213. HOLMAN, R. T., *Progress in the Chemistry of Fats and other Lipids*, Vol. 9, p. 1, (1968) (edited by R. T. Holman, Pergamon Press Ltd.).
214. All references to this numbers deleted.
215. HOLMAN, R. T., *Progress in the Chemistry of Fats and other Lipids*, Vol. 9, p. 275 (1968) (edited by R. T. Holman, Pergamon Press, London).
216. HOLMAN, R. T. and HOFSTETTER, H. H., *J. Amer. Oil Chem. Soc.*, **42,** 540 (1965).
217. HOLMAN, R. T. and RAHM, J. J., *Progress in the Chemistry of Fats and other Lipids*, Vol. 9, p. 13 (1966) (edited by R. T. Holman, Pergamon Press, London).

218. Holub, B. J. and Kuksis, A., *Lipids*, **4,** 466 (1969).
219. Holub, B. J. and Kuksis, A., *J. Lipid Res.*, **12,** 510 (1971).
220. Holub, B. J., Kuksis, A. and Thomson, W., *J. Lipid Res.*, **11,** 558 (1970).
221. Hooghwinkel, G. J. M., Borri, P. and Riemersma, J. C., *Rec. Trav. Chim.*, **83,** 576 (1964).
222. Hooper, N. K. and Law, J. H., *J. Lipid Res.*, **9,** 270 (1968).
223. Hopkins, C. Y., *Progress in the Chemistry of Fats and other Lipids*, Vol. 8, p. 213, (1966) (edited by R. T. Holman, Pergamon Press, London).
224. Hopkins, C. Y., *J. Amer. Oil Chem. Soc.*, **45,** 778 (1968).
225. Hopkins, C. Y. and Bernstein, H. J., *Can. J. Chem.*, **37,** 775 (1959).
226. Hopkins, S. M., Sheehan, G. and Lyman, R. L., *Biochim. Biophys. Acta*, **164,** 272 (1968).
227. Horning, E. C., Karmen, A. and Sweeley, G. C., *Progress in the Chemistry of Fats and other Lipids*, Vol. 7, p. 167 (1964) (edited by R. T. Holman, Pergamon Press, London).
228. Horning, M. G., Casparrini, G. and Horning, E. C., *J. Chromatogr. Sci.*, **7,** 267 (1969).
229. Horning, M. G., Murakami, S. and Horning, E. C., *Amer. J. Clin. Nutr.*, **24,** 1086 (1971).
230. Horrocks, L. A. and Cornwell, D. G., *J. Lipid Res.*, **3,** 165 (1962).
231. Howard, C. F. and Kittinger, G. W., *Lipids*, **2,** 438 (1967).
232. Howard, G. A. and Martin, A. J. P., *Biochem. J.*, **46,** 532 (1950).
233. Huang, A. and Firestone, D., *J. Ass. Off. Anal. Chem.*, **54,** 1288 (1971).
234. Huang, T. C., Chen, C. P., Wefler, V. and Raftery, A., *Anal. Chem.*, **33,** 1405 (1961).
235. IUPAC–IUB Commission on Biochemical Nomenclature, *Eur. J. Biochem.*, **2,** 127 (1967); *Biochem. J.*, **105,** 897 (1967).
236. Iverson, J. L. and Weik, R. W., *J. Ass. Off. Anal. Chem.*, **50,** 1111 (1967).
237. Iyengar, B. T. R. and Schlenk, H., *Biochemistry*, **6,** 396 (1967).
238. James, A. T., *New Biochemical Separations*, p. 2 (1964) (edited by A. T. James and L. J. Morrison, Van Nostrand, London).
239. Jamieson, G. R., *Topics in Lipid Chemistry*, Vol. 1, p. 107 (1970) (edited by F. D. Gunstone, Logos Press, London).
240. Jamieson, G. R. and Reid, E. H., *J. Chromatogr.*, **20,** 232 (1965).
241. Jamieson, G. R. and Reid, E. H., *J. Chromatogr.*, **40,** 160 (1969).
242. Jatzkewitz, H. and Mehl, E., *Hoppe-Seyler's Z. Physiol. Chem.*, **320,** 251 (1960).
243. Jensen, R. G., *Progress in the Chemistry of Fats and other Lipids*, Vol. 11, p. 347 (1971) (edited by R. T. Holman, Pergamon Press, London).
244. Jensen, R. G., Sampugna, J., Carpenter, D. L. and Pitas, R. E., *J. Dairy Sci.*, **50,** 231 (1967).
245. Johnson, A. R., Murray, K. E., Fogerty, A. C., Kennett, B. H., Pearson, J. A. and Shenstone, F. S., *Lipids*, **2,** 316 (1967).
246. Johnson, C. B. and Holman, R. T., *Lipids*, **1,** 371 (1966).
247. Johnson, C. B., Pearson, A. M. and Dugan, L. R., *Lipids*, **5,** 958 (1970).
248. Kanfer, J. N., *Methods in Enzymology*, Vol. XIV, p. 660 (1969) (edited by J. M. Lowenstein, Academic Press, New York).
249. Kapoulas, V. M., *Biochim. Biophys. Acta*, **176,** 324 (1969).
250. Karlsson, K-A., *Acta Chem. Scand.*, **18,** 2395 (1964).
251. Karlsson, K-A., *Acta Chem. Scand.*, **19,** 2425 (1965).

252. KARLSSON, K-A., *Acta Chem. Scand.*, **21,** 2577 (1967).
253. KARLSSON, K-A., *Acta Chem. Scand.*, **22,** 3050 (1968).
254. KARLSSON, K-A., *Chem. Phys. Lipids*, **5,** 6 (1970).
255. KARLSSON, K-A., *Lipids*, **5,** 878 (1970).
256. KARLSSON, K-A. and MARTENSSON, E., *Biochim. Biophys. Acta*, **152,** 230 (1968).
257. KARLSSON, K-A. and PASCHER, I., *J. Lipid Res.*, **12,** 466 (1971).
258. KARMEN, A., *J. Amer. Oil Chem. Soc.*, **44,** 18 (1967).
259. KARMEN, A., *Methods in Enzymology*, Vol. XIV, p. 465 (1969) (edited by J. M. Lowenstein, Academic Press, New York).
260. KARMEN, A. and TRITCH, H. R., *Nature*, **186,** 150 (1960).
261. KATES, M., *Can. J. Biochem. Physiol.*, **34,** 967 (1956).
262. KATES, M., *J. Lipid Res.*, **5,** 132 (1964).
263. KATES, M., *Lipid Chromatographic Analysis*, Vol. 1, p. 1 (1967) (edited by G. V. Marinetti, Edward Arnold Ltd., London).
264. KATES, M., *Advances in Lipid Research*, Vol. 8, p. 225 (1970) (edited by R. Paoletti and D. Kritchevsky, Academic Press, London).
265. KAUFMANN, H. P. and DAS, B., *Fette Seifen Anstrich.*, **64,** 214 (1962).
266. KAUFMANN, H. P., MAKUS, Z. and DIECKE, F., *Fette Seifen Anstrich.*, **63,** 235 (1961).
267. KAUFMANN, H. P., MAKUS, Z. and KHOE, T. H., *Fette Seifen Anstrich.*, **64,** 1 (1962).
268. KAUFMANN, H. P. and VISWANATHAN, C. V., *Fette Seifen Anstrich.*, **65,** 538 (1963).
269. KAUFMANN, H. P. and VISWANATHAN, C. V., *Fette Seifen Anstrich.*, **65,** 925 (1963).
270. KAUFMANN, H. P. and WESSELS, H., *Fette Seifen Anstrich.*, **66,** 81 (1964).
271a. KAULEN, H. D., *Anal. Biochem.*, **45,** 664 (1972).
271b. KEENAN, T. W., AWASTHI, Y. C. and CRANE, F. L., *Biochim. Biophys. Res. Commun.*, **40,** 1102 (1970).
272. KEMP, P. and DAWSON, R. M. C., *Biochem. J.*, **109,** 477 (1968).
273. KENNER, G. W. and STENHAGEN, E., *Acta Chem. Scand.*, **18,** 1551 (1964).
274. KISHIMOTO, Y. and RADIN, N. S., *J. Lipid Res.*, **4,** 130 (1963).
275. KITTREDGE, J. S. and ROBERTS, E., *Science*, **164,** 37 (1969).
276. KLEBE, J. F., FINKBEINER, H. and WHITE, D. M., *J. Amer. Chem. Soc.*, **88,** 3390 (1966).
277. KLEIMAN, R., SPENCER, G. F. and EARLE, F. R., *Lipids*, **4,** 118 (1969).
278. KLENK, E., *Progress in the Chemistry of Fats and other Lipids*, Vol. 10, p. 249 (1970) (edited by R. T. Holman, Pergamon Press, London).
279. KNITTLE, J. L. and HIRSCH, J., *J. Lipid Res.*, **6,** 565 (1965).
280. KOMAREK, R. J., JENSEN, R. G. and PICKETT, B. W., *J. Lipid Res.*, **5,** 268 (1964).
281. KORITALA, S. and ROHWEDDER, W. K., *Lipids*, **7,** 274 (1972).
282. KUHN, R. and WIEGANDT, H., *Chem. Ber.*, **96,** 866 (1963).
283. KUKSIS, A., *Can. J. Biochem.*, **42,** 407 (1964).
284. KUKSIS, A., *Lipid Chromatographic Analysis*, Vol. 1, p. 239 (1967) (edited by G. V. Marinetti, Edward Arnold Ltd., London).
285. KUKSIS, A., *Can. J. Biochem.*, **49,** 1245 (1971).
286. KUKSIS, A., *Fette Seifen Anstrich.*, **73,** 130 (1971).
287. KUKSIS, A., *Fette Seifen Anstrich.*, **73,** 332 (1971).
288. KUKSIS, A., *J. Chromatogr.*, **10,** 53 (1972).
289. KUKSIS, A., *Progress in the Chemistry of Fats and other Lipids*, Vol. 12, p. 1 (1972) (edited by R. T. Holman, Pergamon Press, London).
290. KUKSIS, A. and BRECKENRIDGE, W. C., *J. Lipid Res.*, **7, 576** (1966).
291. KUKSIS, A. and LUDWIG, J., *Lipids*, **1,** 202 (1966).

292. KUKSIS, A. and MARAI, L., *Lipids*, **2**, 217 (1967).
293. KUKSIS, A., STACHNYK, O. and HOLUB, B. J., *J. Lipid Res.*, **10**, 660 (1969).
294. KUSAMRAN, K. and POLGAR, N., *Lipids*, **6**, 961 (1971).
295. KYRIAKIDES, E. C. and BALINT, J. A., *J. Lipid Res.*, **9**, 142 (1968).
296. LANDS, W. E. M. and HART, P., *Biochim. Biophys. Acta* **98**, 532 (1965).
297. LANDS, W. E. M., PIERINGER, R. A., SLAKEY, P. M. and ZSCHOCKE, A., *Lipids*, **1**, 444 (1966).
298. LANDS, W. E. M. and SLAKEY, P. M., *Lipids*, **1**, 295 (1966).
299. LEDEEN, R., *J. Amer. Oil Chem. Soc.*, **43**, 57 (1966).
300. LEDEEN, R., *Chem. Phys. Lipids*, **5**, 205 (1970).
301. LEDEEN, R., SALSMAN, K. and CABRERA, M., *J. Lipid Res.*, **9**, 129 (1968).
302. LEIKOLA, E., NIEMINEN, E. and SOLOMAA, E., *J. Lipid Res.*, **6**, 490 (1965).
303. LEMIEUX, R. U. and VON RUDLOFF, E., *Can. J. Chem.*, **33**, 1701 (1955).
304. LINSTEAD, R. P. and WHALLEY, M., *J. Chem. Soc.*, 2987 (1950).
305. LITCHFIELD, C., *Lipids*, **3**, 170 (1968).
306. LITCHFIELD, C., *Analysis of Triglycerides* (Academic Press, New York) (1972).
307. LITCHFIELD, C., ACKMAN, R. G., SIPOS, J. C. and EATON, C. A., *Lipids*, **6**, 674 (1971).
308. LITCHFIELD, C., HARLOW, R. D. and REISER, R., *J. Amer. Oil Chem. Soc.*, **42**, 849 (1965).
309. LITCHFIELD, C., HARLOW, R. D. and REISER, R., *Lipids*, **2**, 363 (1967).
310. LITCHFIELD, C., REISER, R. and ISBELL, A. F., *J. Amer. Oil Chem. Soc.*, **41**, 52 (1964).
311. LONGONE, D. T. and MILLER, A. H., *Chem. Commun.*, 447 (1967).
312. LOUGH, A. K., *Biochem. J.*, **90**, 4c (1964).
313. LOUGH, A. K., FELINSKI, L. and GARTON, G. A., *J. Lipid Res.*, **3**, 478 (1962).
314. LOWRY, R. R., *J. Lipid Res.*, **9**, 397 (1968).
315. LUCAS, C. C. and RIDOUT, J. H., *Progress in the Chemistry of Fats and other Lipids*, Vol. 10, p. 1 (1970) (edited by R. T. Holman, Pergamon Press, London).
316. LUDDY, F. E., BARFORD, R. A., HERB, S. F., MAGIDMAN, P. and RIEMENSCHNEIDER, F., *J. Amer. Oil Chem. Soc.*, **41**, 603 (1964).
317. LUTHRA, M. G. and SHELTAWY, A., *Biochem. J.*, **126**, 251 (1972).
318. LUTHRA, M. G. and SHELTAWY, A., *Biochem. J.*, **126**, 1231 (1972).
319. McCLOSKEY, J. A., *Methods in Enzymology*, Vol. XIV, p. 382 (1969) (edited by J. M. Lowenstein, Academic Press, New York).
320. McCLOSKEY, J. A., *Topics in Lipid Chemistry*, Vol. 1, p. 369 (1970) (edited by F. D. Gunstone, Logos Press, London).
321. McCLOSKEY, J. A. and LAW, J. H., *Lipids*, **2**, 225 (1967).
322. McCLOSKEY, J. A. and McCLELLAND, M. J., *J. Amer. Chem. Soc.*, **87**, 5090 (1965).
323. McCLUER, R. H., *Chem. Phys. Lipids*, **5**, 220 (1970).
324. MACFARLANE, M. G., *Advances in Lipid Research*, Vol. 2, p. 91 (1964) (edited by R. Paoletti and D. Kritchevsky, Academic Press, London).
325. McKIBBIN, J. M., *Lipid Chromatographic Analysis*, Vol. 1, p. 497 (1967) (edited by G. V. Marinetti, Edward Arnold Ltd., London).
326. McLAFFERTY, F. W., *Anal. Chem.*, **31**, 82 (1959).
327. MAERKER, G., HAEBERER, E. T. and HERB, S. F., *J. Amer. Oil Chem. Soc.*, **43**, 505 (1966).
328. MAHADEVAN, V., *Lipid Chromatographic Analysis*, Vol. 1, p. 191 (1967) (edited by G. V. Marinetti, Edward Arnold Ltd., London).

329. MAHADEVAN, V., *Progress in the Chemistry of Fats and other Lipids*, Vol. 11, p. 81 (1971) (edited by R. T. Holman, Academic Press, London).

330. MAHADEVAN, V., PHILLIPS, F. and LUNDBERG, W. O., *J. Lipid Res.*, **6,** 434 (1965).

331. MAHADEVAN, V., PHILLIPS, F. and LUNDBERG, W. O., *Lipids*, **1,** 183 (1966).

332. MAHADEVAN, V., VISWANATHAN, C. V. and PHILLIPS, F., *J. Lipid Res.*, **8,** 2 (1967).

333. MALINS, D. C., *Progress in the Chemistry of Fats and other Lipids*, Vol. 8, p. 203 (1966) (edited by R. T. Holman, Pergamon Press, London).

334. MALINS, D. C. and MANGOLD, H. K., *J. Amer. Oil Chem. Soc.*, **37,** 576 (1960).

335. MANGOLD, H. K., *Thin Layer Chromatography*, p. 137 (1965) (edited by E. Stahl, Springer-Verlag, Germany).

336. MANGOLD, H. K., *J. Labelled Compounds*, **4,** 2 (1968).

337. MANGOLD, H. K. and BAUMANN, W. J., *Lipid Chromatographic Analysis*, Vol. 1, p. 339 (1967) (edited by G. V. Marinetti, Edward Arnold Ltd., London).

338. MANGOLD, H. K. and SAND, D. M., *Biochim. Biophys. Acta*, **164,** 124 (1968).

339. MARSH, J. B. and WEINSTEIN, D. B., *J. Lipid Res.*, **7,** 574 (1966).

340. MARTENSSON, E., *Progress in the Chemistry of Fats and other Lipids*, Vol. 10, p. 365 (1970) (edited by R. T. Holman, Pergamon Press, London).

341. MEAD, J. F., *Progress in the Chemistry of Fats and other Lipids*, Vol. 9, p. 159 (1968) (edited by R. T. Holman, Pergamon Press, London).

342. MEAD, J. F. and HOWTON, D. R., *Radioisotope Studies of Fatty Acid Metabolism*, p. 119 (1960) (Pergamon Press, London)

343. MEAKINS, G. D. and SWINDELLS, R., *J. Chem. Soc.*, 1044 (1959).

344. MECHAM, D. K. and MOHAMMAD, A., *Cereal Chem.*, **32,** 405 (1955).

345. METCALFE, L. D., *J. Gas Chromatogr.*, **1,** 7 (1963).

346. MIKOLAJCZAK, K. L. and BAGBY, M. O., *J. Amer. Oil Chem. Soc.*, **41,** 391 (1964).

347. MINNIKIN, D. E., *Lipids*, **7,** 398 (1972).

348. MINNIKIN, D. E. and POLGAR, N., *Chem. Commun.*, 312 (1967).

349. MICHELL, R. H., HAWTHORNE, J. N., COLEMAN, R. and KARNOVSKY, M. L., *Biochim. Biophys. Acta*, **210,** 86 (1970).

350. MIWA, T. K., KWOLEK, W. F. and WOLFF, I. A., *Lipids*, **1,** 152 (1966).

351. MIWA, T. K., MIKOLAJCZAK, K. L., EARLE, F. R. and WOLFF, I. A., *Anal. Chem.*, **32,** 1739 (1960).

352. MOLD, J. D., MEANS, R. E., STEVENS, R. K. and RUTH, J. M., *Biochemistry*, **3,** 1293 (1964).

353. MOORE, J. H. and WILLIAMS, D. L., *Biochim. Biophys. Acta*, **84,** 41 (1964).

354. MORELL, P. and BRAUN, P., *J. Lipid Res.*, **13,** 293 (1972).

355. MORRIS, L. J., *J. Chromatogr.*, **12,** 321 (1963).

356. MORRIS, L. J., *J. Lipid Res.*, **7,** 717 (1966).

357. MORRIS, L. J. and HALL, S. W., *Lipids*, **1,** 188 (1966).

358. MORRIS, L. J., Holman, R. T. and FONTELL, K., *J. Lipid Res.*, **1,** 412 (1960).

359. MORRIS, L. J. and MARSHALL, M. O., *Chem. Ind. (London)*, 460 (1966).

360. MORRIS, L. J., MARSHALL, M. O. and KELLY, W., *Tetrahedron Letters*, 4249 (1966).

361. MORRIS, L. J. and WHARRY, D. M., *J. Chromatogr.*, **20,** 27 (1965).

362. MORRIS, L. J., WHARRY, D. M. and HAMMOND, E. W., *J. Chromatogr.*, **31,** 69 (1967).

363. MORRIS, L. J., WHARRY, D. M. and HAMMOND, E. W., *J. Chromatogr.*, **33,** 471 (1968).

364. MORRISON, W. R., *Biochim. Biophys. Acta*, **176,** 537 (1969).
365. MORRISON, W. R., LAWRIE, T. D. V. and BLADES, J., *Chem. Ind. (London)*, 1534 (1961).
366. MORRISON, W. R. and SMITH, L. M., *J. Lipid Res.*, **5,** 600 (1964).
367. MURRAY, K. E., *Austral. J. Chem.*, **12,** 657 (1959).
368. NADENICEK, J. D. and PRIVETT, O. S., *Chem. Phys. Lipids*, **2,** 409 (1968).
369. NELSON, G. J., *Lipids*, **2,** 64 (1967).
370. NELSON, G. J., *Lipids*, **3,** 104 (1968).
371. NEUDOERFFER, T. S. and LEA, C. H., *J. Chromatogr.*, **21,** 138 (1966).
372. NEWMAN, H. A. I., LIU, C. and ZILVERSMIT, D. B., *J. Lipid Res.*, **2,** 403 (1961).
373. NICHOLS, B. W., *Biochim. Biophys. Acta*, **70,** 417 (1963).
374. NICHOLS, B. W., *New Biochemical Separations*, p. 321 (1964) (edited by A. T. James and L. J. Morris, Van Nostrand, New York).
375. NICHOLS, B. W. and James, A. T., *Fette Seifen Anstrich.*, **66,** 1003 (1964).
376. NICHOLS, B. W. and MOORHOUSE, R., *Lipids*, **4,** 311 (1969).
377. NICKELL, E. C. and PRIVETT, O. S., *Separation Sci.*, **2,** 307 (1967).
378. NICOLAIDES, N. and FU, H. C., *Lipids*, **4,** 83 (1969).
379. NICOLOSI, R. J., SMITH S. C. and SANTERRE, R. F., *J. Chromatogr.*, **60,** 111 (1971).
380. NIEHAUS, W. G. and RYHAGE, R., *Anal. Chem.*, **40,** 1840 (1968).
381. NODA, M. and FUJIWARA, N., *Biochim. Biophys. Acta*, **137,** 199 (1967).
382. NUTTER, L. J. and PRIVETT, O. S., *Lipids*, **1,** 234 (1966).
383. NUTTER, L. J. and PRIVETT, O. S., *Lipids*, **1,** 258 (1966).
384. NUTTER, L. J. and PRIVETT, O. S., *J. Dairy Sci.*, **50,** 298 (1967).
385. NUTTER, L. J. and PRIVETT, O. S., *J. Dairy Sci.*, **50,** 1194 (1967).
386. O'BRIEN, J. S. and BENSON, A. A., *J. Lipid Res.*, **5,** 432 (1964).
387. O'CONNOR, R. T., *J. Amer. Oil Chem. Soc.*, **33,** 1 (1956).
388. OKUYAMA, H. and NOJIMA, S., *J. Biochem. (Tokyo)*, **57,** 529 (1965).
389. ORD, W. O. and BAMFORD, P. C., *Chem. Ind. (London)*, 1681 (1966).
390. OTTOLENGHI, A. C., *Methods in Enzymology*, Vol. XIV, p. 188 (1969) (edited by J. M. Lowenstein, Academic Press, New York).
391. OWENS, K., *Biochem. J.*, **100,** 354 (1966).
392. PADLEY, F. B., *Chromatogr. Rev.*, **8,** 208 (1966).
393. PALMER, F. B. St. C., *Biochim. Biophys. Acta*, **231,** 134 (1971).
394. PARKER, F. and PETERSON, N. F., *J. Lipid Res.*, **6,** 455 (1965).
395. PEIFER, J. J., *Mikrochim. Acta*, 529 (1962).
396. PENICK, R. J., MEISLER, M. H. and McCLUER, R. H., *Biochim. Biophys. Acta*, **116,** 279 (1966).
397. PITT, G. A. J. and MORTON, R. A., *Progress in the Chemistry of Fats and other Lipids*, Vol. 4, p. 228 (1957) (edited by R. T. Holman, Pergamon Press, London).
398. POHL, P., GLASL, H. and WAGNER, H., *J. Chromatogr.*, **49,** 488 (1970).
399. POLGAR, N., *Topics in Lipid Chemistry*, Vol. 2, p. 207 (1971) (edited by F. D. Gunstone, Logos Press, London).
400. POLITO, A. J., AKITA, T. and SWEELEY, C. C., *Biochemistry*, **7,** 2609 (1968).
401. POLITO, A. J., NAWORAL, J. and SWEELEY, C. C., *Biochemistry*, **8,** 1811 (1969).
402. POSSMAYER, F., SCHERPHOF, G. L., DUBBELMAN, T. M. A. R., VAN GOLDE, L. M. G. and VAN DEENEN, L. L. M., *Biochim. Biophys. Acta*, **176,** 95 (1969).
403. POWELL, R. G. and SMITH, C. R., *Biochemistry*, **5,** 625 (1966).
404. POWELL, R. G., SMITH, C. R., GLASS, C. A. and WOLFF, I. A., *J. Org. Chem.*, **30,** 610 (1965).

320   *References*

405. Privett, O. S., *Progress in the Chemistry of Fats and other Lipids*, Vol. 9, p. 91 (1966) (edited by R. T. Holman, Pergamon Press, London).

406. Privett, O. S., *Progress in the Chemistry of Fats and other Lipids*, Vol. 9, p. 407 (1966) (edited by R. T. Holman, Pergamon Press, London).

407. Privett, O. S. and Blank, M. L., *J. Amer. Oil Chem. Soc.*, **40**, 70 (1963).

408. Privett, O. S., Dougherty, K. A. and Castell, J. D., *Amer. J. Clin. Nutr.*, **24**, 1265 (1971).

409. Privett, O. S., Nadenicek, J. D., Weber, R. P. and Pusch, F., *J. Amer. Oil Chem. Soc.*, **40**, 28 (1963).

410. Privett, O. S. and Nickell, E. C., *J. Amer. Oil Chem. Soc.*, **40**, 189 (1963).

411. Privett, O. S. and Nickell, E. C., *Lipids*, **1**, 98 (1966).

412. Privett, O. S. and Nutter, L. J., *Lipids*, **2**, 149 (1967).

413. Purcell, J. E. and Ettre, L. S., *J. Gas Chromatogr.*, **4**, 23 (1966).

414. Radin, N. S., *J. Amer. Oil Chem. Soc.*, **42**, 569 (1965).

415. Raju, P. K. and Reiser, R., *Lipids*, **1**, 10 (1966).

416. Raju, P. K. and Reiser, R., *Lipids*, **2**, 197 (1967).

417. Ramachandran, S., Panganamala, R. V. and Cornwell, D. G., *J. Lipid Res.*, **10**, 465 (1969).

418. Ramachandran, S., Sprecher, H. W. and Cornwell, D. G., *Lipids*, **3**, 511 (1968).

419. Ramachandran, S., Venkata Rao, P. and Cornwell, D. G., *J. Lipid Res.*, **9**, 137 (1968).

420. Ramwell, P. W., Shaw, J. E., Clarke, G. B., Grostic, M. F., Kaiser, D. G. and Pike, K. E., *Progress in the Chemistry of Fats and other Lipids*, Vol. 9, p. 231 (1968) (edited by R. T. Holman, Pergamon Press, London).

421. Rapport, M. M. and Alonzo, N., *J. Biol. Chem.*, **217**, 199 (1955).

422. Recourt, J. H., Jurriens, G. and Schmitz, M., *J. Chromatogr.*, **30**, 35 (1967).

423. Renkonen, O., *Acta Chem. Scand.*, **16**, 1288 (1962).

424. Renkonen, O., *Acta Chem. Scand.*, **17**, 275, 634 and 1925 (1963).

425. Renkonen, O., *J. Amer. Oil Chem. Soc.*, **42**, 298 (1965).

426. Renkonen, O., *Ann. Med. Exptl. Biol. Fenniae (Helsinki)*, **44**, 356 (1966).

427. Renkonen, O., *Lipids*, **1**, 160 (1966).

428. Renkonen, O., *Biochim. Biophys. Acta*, **125**, 288 (1966).

429. Renkonen, O., *Biochim. Biophys. Acta*, **137**, 575 (1967).

430. Renkonen, O., *Advances in Lipid Research*, Vol. 5, p. 329 (1967) (edited by R. Paoletti and D. Kritchevsky, Academic Press, New York).

431. Renkonen, O., *Acta Chem. Scand.*, **21**, 1108 (1967).

432. Renkonen, O., *J. Lipid Res.*, **9**, 34 (1968).

433. Renkonen, O., *Biochim. Biophys. Acta*, **152**, 114 (1968).

434. Renkonen, O., *Lipids*, **3**, 191 (1968).

435. Renkonen, O. and Hirvisalo, E. L., *Biochem. J.*, **89**, 29P (1963).

436. Renkonen, O., Luisvaara, S. and Miettinen, E., *Ann. Med. Exptl. Biol. Fenniae (Helsinki)*, **43**, 200 (1965).

437. Renkonen, O. and Varo, P., *Lipid Chromatographic Analysis*, Vol. 1, p. 41 (1967) (edited by G. V. Marinetti, Edward Arnold Ltd., London).

438. Roberts, R. N., *Lipid Chromatographic Analysis*, Vol. 1, p. 447 (1967) (edited by G. V. Marinetti, Edward Arnold Ltd., London).

439. Robertson, A. F. and Lands, W. E. M., *Biochemistry*, **1**, 804 (1962).

440. Roch, L. A. and Grossberg, S. E., *Anal. Biochem.*, **41**, 105 (1971).

441. Roehm, J. N. and Privett, O. S., *Lipids*, **5**, 353 (1970).

442. ROUSER, G., BAUMAN, A. J., KRITCHEVSKY, G., HELLER, D. and O'BRIEN, J., *J. Amer. Oil Chem. Soc.*, **38**, 554 (1961).
443. ROUSER, G., GALLI, C., LIEBER, E., BLANK, M. L. and PRIVETT, O. S., *J. Amer. Oil Chem. Soc.*, **41**, 836 (1964).
444. ROUSER, G., KRITCHEVSKY, G., HELLER, D. and LIEBER, E., *J. Amer. Oil Chem. Soc.*, **40**, 425 (1963).
445. ROUSER, G., KRITCHEVSKY, G., SIMON, G. and NELSON, G. J., *Lipids*, **2**, 37 (1967).
446. ROUSER, G., KRITCHEVSKY, G. and YAMAMOTO, A., *Lipid Chromatographic Analysis*, Vol. 1, p. 99 (1967) (edited by G. V. Marinetti, Edward Arnold Ltd., London).
447. ROUSER, G., KRITCHEVSKY, G., YAMAMOTO, A., SIMON, G., GALLI, C. and BAUMAN, A. J., *Methods in Enzymology*, Vol. XIV, p. 272 (1969) (edited by J. M. Lowenstein, Academic Press, USA).
448. ROUSER, G., O'BRIEN, J. and HELLER, D., *J. Amer. Oil Chem. Soc.*, **38**, 14 (1961).
449. RUMSBY, M. G., *J. Chromatogr.*, **42**, 237 (1969).
450. RYHAGE, R. and STENHAGEN, E., *Arkiv Kemi*, **13**, 523 (1959).
451. RYHAGE, R. and STENHAGEN, E., *Arkiv Kemi*, **15**, 291 (1960).
452. RYHAGE, R. and STENHAGEN, E., *Arkiv Kemi*, **15**, 333 (1960).
453. RYHAGE, R. and STENHAGEN, E., *Arkiv Kemi*, **15**, 545 (1960).
454. SAFFORD, R. and NICHOLS, B. W., *Biochim. Biophys. Acta*, **210**, 57 (1970).
455. SAITO, T. and HAKOMORI, S., *J. Lipid Res.*, **12**, 257 (1971).
456. SAMUELSSON, B. and SAMUELSSON, K., *Biochim. Biophys. Acta*, **164**, 421 (1968).
457. SAMUELSSON, B. and SAMUELSSON, K., *J. Lipid Research*, **10**, 41 (1969).
458. SANDERS, H., *Biochim. Biophys. Acta*, **144**, 485 (1967).
459. SCHIEFER, H-G. and NEUHOFF, V., *Hoppe-Seyler's Z. Physiol. Chem.*, **352**, 913 (1971).
460. SCHLENK, H., *Progress in the Chemistry of Fats and other Lipids*, Vol. 2, p. 243 (1954) (edited by R. T. Holman, Pergamon Press, London).
461. SCHLENK, H. and GELLERMAN, J. L., *Anal. Chem.*, **32**, 1412 (1960).
462. SCHLENK, H. and GELLERMAN, J. L., *J. Amer. Oil Chem. Soc.*, **38**, 555 (1961).
463. SCHLENK, H. and SAND, D. M., *Anal. Chem.*, **34**, 1676 (1962).
464. SCHMID, H. H. O., BAUMANN, W. J. and MANGOLD, H. K., *J. Amer. Chem. Soc.*, **89**, 4797 (1967).
465. SCHMID, H. H. O., JONES, L. L. and MANGOLD, H. K., *J. Lipid Res.*, **8**, 692 (1967).
466. SCHMID, H. H. O. and MANGOLD, H. K., *Biochim. Biophys. Acta*, **125**, 182 (1966).
467. SCHMID, H. H. O. and MANGOLD, H. K., *Biochem. Zeits.*, **346**, 13 (1966).
468. SCHMID, H. H. O., MANGOLD, H. K. and LUNDBERG, W. O., *J. Amer. Oil Chem. Soc.*, **42**, 372 (1965).
469. SCHMID, H. H. O., MANGOLD, H. K., LUNDBERG, W. O. and BAUMANN, W. J., *Microchem. J.*, **11**, 306 (1966).
470. SCHMID, H. H. O. and TAKAHASHI, T., *Hoppe-Seyler's Z. Physiol. Chem.*, **349**, 1673 (1968).
471. SCHNEIDER, P. B., *J. Lipid Res.*, **7**, 169 (1966).
472. SCHOGT, J. C. M. and HAVERKAMP-BEGEMANN, P., *J. Lipid Res.*, **6**, 466 (1965).
473. SCHOLFIELD, C. R. and DUTTON, H. J., *J. Amer. Oil Chem. Soc.*, **47**, 1 (1970).
474. SCHOOR, W. P. and MELIUS, P., *Biochim. Biophys. Acta*, **187**, 186 (1969).
475. SCHULTE, K. E. and RUCKER, G., *J. Chromatogr.*, **49**, 317 (1970).
476. SCOTT, P. G. W., *Process Biochem.*, 16 (1967).
477. SERDAREVICH, B. and CARROLL, K. K., *Can. J. Biochem.*, **44**, 743 (1966).
478. SERDAREVICH, B. and CARROLL, K. K., *J. Lipid Res.*, **7**, 277 (1966).

479. SHAW, N., *Biochim. Biophys. Acta*, **164**, 435 (1968).
480. SHAW, N., *Bacteriol. Revs.*, **34**, 365 (1970).
481. SHAW, W. A. and HARLAN, W. R., *Anal. Biochem.*, **43**, 119 (1971).
482. SIAKOTOS, A. N., *Lipids*, **2**, 87 (1967).
483. SIAKOTOS, A. N. and ROUSER, G., *J. Amer. Oil Chem. Soc.*, **42**, 913 (1965).
484. SIMPSON, T. H., *J. Chromatogr.*, **38**, 24 (1968).
485. SINGH, H. and PRIVETT, O. S., *Lipids*, **5**, 692 (1970).
486. SINGLETON, W. S., GRAY, M. S., BROWN, M. L. and WHITE, J. L., *J. Amer. Oil Chem. Soc.*, **42**, 53 (1965).
487. SINNHUBER, R. O., CASTELL, J. D. and LEE, D. J., *Fed. Proc. Fed. Amer. Soc. Exp. Biol.*, **31**, 1436 (1972).
488. SKIDMORE, W. D. and ENTENMAN, C., *J. Lipid Res.*, **3**, 471 (1962).
489. SKIPSKI, V. P. and BARCLAY, M., *Methods in Enzymology*, Vol. XIV, p. 530 (1969) (edited by J. M. Lowenstein, Academic Press, New York).
490. SKIPSKI, V. P., BARCLAY, M., REICHMAN, E. S. and GOOD, J. J., *Biochim. Biophys. Acta*, **137**, 80 (1967).
491. SKIPSKI, V. P., PETERSON, R. F. and BARCLAY, M., *Biochem. J.*, **90**, 374 (1964).
492. SKIPSKI, V. P., SMOLOWE, A. F. and BARCLAY, M., *J. Lipid Res.*, **8**, 295 (1967).
493. SLIPSKI, V. P., SMOLOWE, A. F., SULLIVAN, R. C. and BARCLAY, M., *Biochim. Biophys. Acta*, **106**, 386 (1965).
494. SLAKEY, P. M. and LANDS, W. E. M., *Lipids*, **3**, 30 (1968).
495. SLAWSON, V. and MEAD, J. F., *J. Lipid Res.*, **13**, 143 (1972).
496. SLAWSON, V. and STEIN, R. A., *Lipids*, **5**, 713 (1970).
497. SLOTBOOM, A. J., DE HAAS, G. H., BONSEN, P. P. M., BURBACH-WESTERHUIS, G. J. and VAN DEENEN, L. L. M., *Chem. Phys. Lipids*, **4**, 15 (1970).
498. SLOTBOOM, A. J., DE HAAS, G. H., BURBACH-WESTERHUIS, G. J. and VAN DEENEN, L. L. M., *Chem. Phys. Lipids*, **4**, 30 (1970).
499. SLOTBOOM, A. J., DE HAAS, G. H. and VAN DEENEN, L. L. M., *Chem. Phys. Lipids*, **1**, 192 (1967).
500. SMITH, C. R., *Lipids*, **1**, 268 (1966).
501. SMITH, C. R., *Progress in the Chemistry of Fats and other Lipids*, Vol. 11, p. 137 (1970) (edited by R. T. Holman, Pergamon Press, London).
502. SMITH, C. R., *Topics in Lipid Chemistry*, Vol. 1, p. 277 (1970) (edited by F. D. Gunstone, Logos Press, London).
503. SMITH, C. R., *Topics in Lipid Chemistry*, Vol. 3, p. 89 (1972) (edited by F. D. Gunstone, Logos Press, London).
504. SMITH, E. D. and SHEPPARD, H., *Nature*, **208**, 878 (1965).
505. SNYDER, F., *Progress in the Chemistry of Fats and other Lipids*, Vol. 10, p. 287 (1970) (edited by R. T. Holman, Pergamon Press, London).
506. SNYDER, F. and KIMBLE, H., *Anal. Biochem.*, **11**, 510 (1965).
507. SNYDER, F. and PIANTADOSI, C., *Advances in Lipid Research*, Vol. 4, p. 257 (1966) (edited by R. Paoletti and D. Kritchevsky, Academic Press, London).
508. SNYDER, F. and PIANTADOSI, C., *Biochim. Biophys. Acta*, **152**, 794 (1968).
509. SNYDER, F. and SMITH, D., *Separation Sci.*, **1**, 709 (1966).
510. SNYDER, W. R. and LAW, J. H., *Lipids*, **5**, 800 (1970).
511. SORM, F., WOLLRAB, V., JAROLIMEK, P. and STREIBL, M., *Chem. Ind. (London)*, 1833 (1964).
512. SPENCE, M. W., *Can. J. Biochem.*, **47**, 735 (1969).
513. SPENCER, B., *Biochem. J.*, **75**, 435 (1960).

514. STEARNS, E. M. and QUACKENBUSH, F. W., Paper presented at 38th Fall Meeting of the American Oil Chemists' Society, Chicago (1964).
515. STEARNS, E. M., WHITE, H. B. and QUACKENBUSH, F. W., *J. Amer. Oil Chem. Soc.*, **39,** 61 (1962).
516. STEIN, R. A. and SLAWSON, V., *J. Chromatogr.*, **25,** 204 (1966).
517. STEIN, R. A. and SLAWSON, V., *Progress in the Chemistry of Fats and other Lipids*, Vol. 8, p. 375 (1966) (edited by R. T. Holman, Pergamon Press, London).
518. STEINBERG, G., SLATON, W. H., HOWTON, D. R. and MEAD, J. F., *J. Biol. Chem.*, **220,** 257 (1956).
519. STODOLA, F. H., DEINEMA, M. H. and SPENCER, J. F. T., *Bacteriol. Revs.*, **31,** 194 (1967).
520. STORRY, J. E. and TUCKLEY, B., *Lipids*, **2,** 501 (1967).
521. STRONG, C., DILS, R. and GALLIARD, T., *Column*, (Pye-Unicam, Ltd.), No. 13, p. 2 (1971).
522. SUN, G. Y. and HORROCKS, L. A., *J. Lipid Res.*, **10,** 153 (1969).
523. SUPELCO, INC., *Chromatography of Lipids*, Vol. 4, No. 2 (1970) published by Supelco Inc., U.S.A.
524. SUPINA, W. R., KRUPPA, R. F. and HENLY, R. S., *J. Amer. Oil Chem. Soc.*, **44,** 74 (1967).
525. SVENNERHOLM, E. and SVENNERHOLM, L., *Biochim. Biophys. Acta*, **70,** 432 (1963).
526. SVENNERHOLM, L., *J. Neurochem.*, **1,** 42 (1956).
527. SVENNERHOLM, L., *Acta Soc. Med. Upsalien*, **61,** 287 (1956).
528. SVENNERHOLM, L., *Biochim. Biophys. Acta*, **24,** 604 (1957).
529. SWEELEY, C. C., *Methods in Enzymology*, Vol. XIV, p. 255 (1969) (edited by J. M. Lowenstein, Academic Press, U.S.A.).
530. SWEELEY, C. C. and MOSCATELLI, E. A., *J. Lipid Res.*, **1,** 40 (1960).
531. SWEELEY, C. C. and VANCE, D. E., *Lipid Chromatographic Analysis*, Vol. 1. p. 465 (1967) (edited by G. V. Marinetti, Edward Arnold Ltd.).
532. SWEELEY, C. C. and WALKER, B., *Anal. Chem.*, **36,** 1461 (1964).
533. TAKAHASHI, T. and SCHMID, H. H. O., *Chem. Phys. Lipids*, **2,** 220 (1968).
534. Tentative Method Cd 14–61 *Official and Tentative Methods of the American Oil Chemists' Society*, (1946) (A.O.C.S. Chicago, U.S.A.).
535. THOMAS, A. E., SHAROUN, J. E. and RALSTON, H., *J. Amer. Oil Chem. Soc.*, **42,** 789 (1965).
536. THOMAS, P. J. and DUTTON, H. J., *Anal. Chem.*, **41,** 657 (1969).
537. THOMPSON, G. A. and LEE, P., *Biochim. Biophys. Acta*, **98,** 151 (1965).
538. THOMPSON, W., *Biochim. Biophys. Acta*, **187,** 150 (1969).
539. THORPE, C. W., POHLAND, L. and FIRESTONE, D., *J. Ass. Off. Anal. Chem.*, **52,** 774 (1969).
540. TINOCO, J., BABCOCK, R., McINTOSH, D. J. and LYMAN, R. L., *Biochim. Biophys. Acta*, **164,** 129 (1968).
541. TRENNER, N. R., SPETH, O. C., GRUBER, V. B. and VANDENHEUVEL, W. J. A., *J. Chromatogr.*, **71,** 415 (1972).
542. TULLOCH, A. P., *J. Amer. Oil Chem. Soc.*, **41,** 833 (1964).
543. TULLOCH, A. P., *J. Amer. Oil Chem. Soc.*, **43,** 670 (1966).
544. TURNER, J. D. and ROUSER, G., *Anal. Biochem.*, **38,** 423 (1970).
545. TURNER, J. D. and ROUSER, G., *Anal. Biochem.*, **38,** 437 (1970).
546. VAN BEERS, G. J., DE JONGH, H. and BOLDINGH, J., *Essential Fatty Acids*, p. 43 (1958) (edited by H. Sinclair, Butterworth Press, London).

547. VAN DEENEN, L. L. M. and DE HAAS, G. H., *Advances in Lipid Research*, Vol. 2, p. 168 (1964) (edited by R. Paoletti and D. Kritchevsky, Academic Press, London).

548. VAN DER HORST, D. J., VAN GENNIP, A. H. and VOOGT, P. A., *Lipids*, **4**, 300 (1969).

549. VAN GOLDE, L. M. G. and VAN DEENEN, L. L. M., *Biochim. Biophys. Acta*, **125**, 496 (1966).

550. VAN GOLDE, L. M. G. and VAN DEENEN, L. L. M., *Chem. Phys. Lipids*, **1**, 157 (1967).

551. VASKOVSKY, V. E. and SUPPES, Z. S., *J. Chromatogr.*, **63**, 455 (1971).

552. VASKOVSKY, V. E. and SVETASHEV, V. I., *J. Chromatogr.*, **65**, 451 (1972).

553. VAUGHAN, M., *J. Biol. Chem.*, **237**, 3354 (1962).

554. VENKATA RAO, P., RAMACHANDRAN, S. and CORNWELL, D. G., *J. Lipid Res.*, **8**, 380 (1967).

555. VERESHCHAGIN, A. G., *J. Chromatogr.*, **14**, 184 (1964).

556. VERGER, R., DE HAAS, G. H., SARDA, L. and DESNUELLE, P., *Biochim. Biophys. Acta*, **188**, 272 (1969).

557. VIOQUE, E. and HOLMAN, R. T., *Anal. Chem.*, **33**, 1444 (1961).

558. VIOQUE, E. and HOLMAN, R. T., *J. Amer. Oil Chem. Soc.*, **39**, 63 (1962).

559. VISWANATHAN, C. V., *Chromatogr. Rev.*, **10**, 18 (1968).

560. VISWANATHAN, C. V., *Chromatogr. Rev.*, **11**, 153 (1969).

561. VISWANATHAN, C. V., HOEVET, S. P., LUNDBERG, W. O., WHITE, J. M. and MUCCINI, G. A., *J. Chromatogr.*, **40**, 225 (1969).

562. VOGEL, A. I., *Practical Organic Chemistry*, (1956) (3rd edition, Longmans-Green, London).

563. VON GORKOM, M. and HALL, G. E., *Spectrochim. Acta*, **22**, 990 (1966).

564. VON RUDLOFF, E., *J. Amer. Oil Chem. Soc.*, **33**, 125 (1956); *Can. J. Chem.*, **34**, 1413 (1956).

565. VORBECK, M. L. and MARINETTI, G. V., *J. Lipid Res.*, **6**, 3 (1965).

566. WAGNER, H., HORHAMMER, L. and WOLFF, P., *Biochem. Z.*, **334**, 175 (1961).

567. WAKU, K. and NAKAZAWA, Y., *J. Biochem. (Tokyo)*, **72**, 149 (1972).

568. WARNER, H. R. and LANDS, W. E. M., *J. Amer. Chem. Soc.*, **85**, 60 (1963).

569. WATSON, G. R. and WILLIAMS, J. P., *J. Chromatogr.*, **67**, 221 (1972).

570. WAYS, P. and HANAHAN, D. J., *J. Lipid Res.*, **5**, 318 (1964).

571. WEBB, R. A. and METTRICK, D. F., *J. Chromatogr.*, **67**, 75 (1972).

572. WEISS, B., *Lipid Chromatographic Analysis*, Vol. 1, p. 429 (1967) (edited by G. V. Marinetti, Edward Arnold Ltd., London).

573. WELLS, M. A. and DITTMER, J. C., *Biochemistry*, **2**, 1259 (1963).

574. WELLS, M. A. and DITTMER, J. C., *J. Chromatogr.*, **18**, 503 (1965).

575. WELLS, M. A. and DITTMER, J. C., *Biochemistry*, **4**, 2459 (1965).

576. WELLS, W. W., PITTMAN, T. A. and WELLS, H. J., *Anal. Biochem.*, **10**, 450 (1965).

577. WHITCUTT, J. M. and SUTTON, D. A., *Biochem. J.*, **63**, 469 (1956).

578. WHITE, H. B. and QUACKENBUSH, F. W., *J. Amer. Oil Chem. Soc.*, **39**, 511 (1962).

579. WHITE, H. B. and QUACKENBUSH, F. W., *J. Amer. Oil Chem. Soc.*, **39**, 517 (1962).

580. WHITTAKER, V. P. and WIJESUNDERA, S., *Biochem. J.*, **51**, 384 (1952).

581. WHITTLE, K. J., DUNPHY, P. J. and PENNOCK, J. F., *Chem. Ind. (London)*, 1303 (1966).

582. WIEGANDT, H., *Advances in Lipid Research*, Vol. 9, p. 249 (1971) (edited by R. Paoletti and D. Kritchevsky, Academic Press, London).

583. WILLNER, D., *Chem. Ind. (London)*, 1839 (1965).
584. WINDELER, A. S. and FELDMAN, G. L., *Lipids*, **4,** 167 (1969).
585. WOLFF, I. A. and MIWA, T. K., *J. Amer. Oil Chem. Soc.*, **42,** 208 (1965).
586. WOOD, P. D. S. and HOLTON, S., *Proc. Soc. Exp. Biol. Med.*, **115,** 990 (1964).
587. WOOD, R., *Lipids*, **2,** 199 (1967).
588. WOOD, R., BEVER, E. L., and SNYDER, F., *Lipids*, **1,** 399 (1966).
589. WOOD, R. and HARLOW, R. D., *J. Lipid Res.*, **10,** 463 (1969).
590. WOOD, R. and HARLOW, R. D., *Arch. Biochem. Biophys.*, **131,** 495 (1969).
591. WOOD, R. and HARLOW, R. D., *Arch. Biochem. Biophys.*, **135,** 272 (1969).
592. WOOD, R. and HEALY, K., *Lipids*, **5,** 661 (1970).
593. WOOD, R. D., RAJU, P. K. and REISER, R., *J. Amer. Oil Chem. Soc.*, **42,** 161 (1965).
594. WOOD, R. and SNYDER, F., *Lipids*, **1,** 62 (1966).
595. WOOD, R. and SNYDER, F., *Lipids*, **2,** 161 (1967).
596. WOOD, R. and SNYDER, F., *Lipids*, **3,** 129 (1968).
597. WOOD, R. and SNYDER, F., *Arch. Biochem. Biophys.*, **131,** 478 (1969).
598. WOODFORD, F. P. and VAN GENT, C. M., *J. Lipid Res.*, **1,** 188 (1960).
599. WREN, J. J., *J. Chromatogr.*, **4,** 173 (1960).
600. WREN, J. J. and SZCZEPANOWSKA, A. D., *J. Chromatogr.*, **14,** 405 (1964).
601. WURSTER, C. F. and COPENHAVER, J. H., *Lipids*, **1,** 422 (1966).
602. WURSTER, C. F., COPENHAVER, J. H. and SCHAFER, P. R., *J. Amer. Oil Chem. Soc.*, **40,** 513 (1963).
603. WUTHIER, R. E., *J. Lipid Res.*, **7,** 558 (1966).
604. WYBENGA, D. R., PILEGGI, V. J., DIRSTINE, P. H. and DI GIORGIO, J., *Clin. Chem.*, **16,** 980 (1970).
605. YABUUCHI, H. and O'BRIEN, J. S., *J. Lipid Res.*, **9,** 65 (1968).
606. YAMAMOTO, A. and ROUSER, G., *Lipids*, **5,** 442 (1970).
607. YANG, H. and HAKOMORI, S., *J. Biol. Chem.*, **246,** 1192 (1971).
608. YOUNG, O. M. and KANFER, J. N., *J. Chromatogr.*, **19,** 611 (1965).
609. YURKOWSKI, M. and BROCKERHOFF, H., *Biochim. Biophys. Acta*, **125,** 55 (1966).
610. ZEMAN, I., *J. Gas Chromatogr.*, **3,** 18 (1965).
611. ZEMAN, I. and POKORNY, J., *J. Chromatogr.*, **10,** 15 (1963).
612. ZINKEL, D. F. and ROWE, J. W., *Anal. Chem.*, **36,** 1160 (1964).

Index

327